The

Rise and Fall

of

Practically Everything

Bill Keith

The Rise and Fall of Practically Everything

"Books for the Journey"
StoneGate Publishing Company, Inc.
Longview 2011

To the culture warriors – heroes all –
standing tall for freedom.

Contents

Preface

A "winner takes all" culture war is raging throughout the United States – a war for the heart and soul of the American people.

The Rise and Fall of Practically Everything is a report from the front lines of the war and carefully chronicles the cataclysmic paradigm shift that has put this nation in peril. I believe we face the serious threat of losing our once-great civilization as government, the news media, our other great institutions, and even the Christian church reel under the assault of a cultural revolution.

The high standards that made this nation great are in decline, yet government seems unable to respond to the burning issues of our time such as the threat from Muslim terrorists and the millions of illegal immigrants living in our land. Our great educational institutions have experienced an ethical meltdown as their leaders embrace a worldview of moral relativism without moral parameters. Left-leaning politicians and movie stars are setting the moral agenda and leading this nation into a hedonistic, X-rated culture. And Christians hide behind stained-glass windows and under the pews unwilling to confront today's culture of decay.

Rise and Fall explores and confronts today's front-page concerns:

- The challenge of countering the paradigm shift that is transforming Christianity into a "single-digit" minority.
- Why perversion and immorality are accepted, but

core American and biblical values are criticized as old-fashioned and bigoted.

- The strategies so-called "secular progressives" have used to drive religion out of the Public Square and Christians to the back of the bus.
- The impact of the eclectic new-age religion promoted by Oprah Winfrey and other celebrities.
- Why today's Christian church in America has such little "salt" and "light" influence in our culture.

This book explains how the secular progressives are leading this once-great civilization toward a dangerous rendezvous with history.

But there is some hope: tens of millions of concerned Americans need not stand by impotently and watch the United States become yet another great nation that failed. There is time to reverse the trend. Men and women of valor can fight back successfully and win this war.

Chapter One
Radical Winds of Change

When you take a close look at all the nonsense in this "land of the free" and "home of the brave," you may think the clowns have taken over the circus. Even the casual observer must recognize that there is a lot of foolishness trying to masquerade as public policy, science, religion, education and a dozen other erstwhile cornerstones of our free and democratic society.

For instance, the men and women in the U. S. Congress are the canonization of mediocrity and masters of the Shell Game on the Potomac. Someone should send all of them copies of *Politics for Dummies* for passing bills they haven't even read.

They dance to the "Beltway Boogie" but won't admit that our land is in financial jeopardy because of run-away spending and, hence, have no desire to fix it. Evidence of something that works and is good for the American people is like a silver bullet to them.

I have come to believe that the our Curly-Larry-and-Moe congressmen are no longer our friends for they are living in a dream world that has become the American nightmare. They assume they can dictate to "We the People" while ignoring our wishes.

Congressmen and women are like phantoms on Capitol Hill for they generally work only two days a week.

We have a former vice president running around the world like Wile E. Coyote with his desperate vision of a-pocalypse warning us that carbon dioxide emissions from our automobiles will burn up the earth within ten years.

Hollywood is a fantasy world of narcissism and the cockpit of self love where a pop tart can go to jail for driving under the influence and emerge a few days later carrying a Bible and acting like Mother Teresa. It's the greatest show on earth and these ringmasters and clowns are setting the social and moral agenda and leading this nation into a hedonistic, X-rated culture.

The motion picture industry's leaders have become the keepers of the American vulgarity, the supporters of counterculture heroes. Most of today's movies are decadent and anyone with a hint of conscience must recoil in disgust at the avalanche of explicit sex, dirty language and sick violence where young girls traffic on their sexuality in a celebrity-driven nation that rewards bad behavior.

But movie stars are considered "cool."

We live in a land where bizarre behavior is so deeply entrenched some people think it's normal, where there are those who believe God is dead but Elvis is alive.

Through the years the members of the U. S. Supreme Court read the Constitution like the funny papers. During the so-called "Warren Court" – 1953-1979 – I thought the justices must be doing some kind of fertility dance for they ruled that nearly any perversion is protected by the First Amendment. Nine old men in black robes ripped this culture apart for they never understood the unforgiving nature of reality. They should have been on the Most Wanted list with their pictures in the post office.

The Warren Court represented the highest level of the rot and decay of the American judicial system with its frantic effort to eradicate all traces of religion, particularly Christianity, from our society. Sometimes the justices acted like they were from another mental galaxy for they tried to define deviancy down and call it normal rather than dealing with the pathologies of fallen human nature and the spiritual poison that infects the human race.

They squeezed prayer and the Ten Commandments out of the schools but allowed sex education programs that teach young people all forms of sexual experimentation. The high court determined that just about anything goes except religious expression.

Most of our states – with the exception of New England and the West Coast – have tried to get rid of pornography. But federal judges say the Constitution guarantees a man's right to publish or read anything he desires. When any of us speak out against this perversion, the TV anchors and editorial writers in the major newspapers call us extremists and bigots.

Is that the America we want to live in? I don't think so.

During the Warren Court years the people of this land witnessed the beginning of the end of the Constitution.

While we languish in nihilism and hedonism, there are church groups hiding behind their stained glass windows or under the pews and surface only long enough to debate whether to ordain homosexuals and lesbians to the gospel ministry.

The pastors of some churches are silent on the burning issues of our time. For thirty-five years they have sat back and done nothing except give lip service to the sanctity of life movement. America's greatest pulpits are virtually silent on the issue.

Although we're fighting a cultural war for the soul of America, most Christians haven't even reported for duty. Perhaps they are secret agents or believe in moral disarmament and have already raised the white flag of surrender. There should be genuine outrage because of what is taking place in our culture but most Christians have been asleep so long they don't know how to take a stand for anything.

Today's church, unable or unwilling to face the challenge of a culture that is crumbling all around us, also has become the target of in-your-face atheists as atheism is "in" and Christianity is "out," particularly with the TV anchors and editorial writers of major newspapers.

Oprah Winfrey, like the Pied Piper of Hamelin, is leading millions into her eclectic New Age religion that includes a heavy emphasis on eastern mysticism and it looks
to me like she and her followers may all go off the cliff together.

Mother Nature has replaced Father God and the worship of the environment has replaced the church in the
hearts and minds of millions who worship at the pagan al-
tars.

One of our most recent presidents was a master of disguise and convoluted rhetoric as he led our nation into an
age of moral insignificance. And the former first lady
walked the halls of the White House talking to the ghosts
of Eleanor Roosevelt and Mahatma Gandhi under the tu-
telage of a well-known Houston, Texas psychic.

Those were real Shangri-la moments!

The standards that made this nation great are being dis-
mantled one at a time as we rush toward an uncertain rendezvous with history.

During the past half-century I have been an eyewitness
to the rise and fall of practically everything in this land and
have come to the conclusion that this is a nation at war
with itself. Even the most casual observer must realize we
face a precarious future. Yet no one seems to know what to
do about it.

We are unable to deal with the burning issues of our
time such as run-away governmental spending, terrorism,
illegal immigrants, health care, a staggering economy,
double-digit unemployment and the decline in religion.

Islamic radicals, a world-wide, rapidly growing band of
terrorists, are obsessed with destroying our land. Although
these terrorists call the United States the "Great Satan" and
have killed thousands of our people in New York City,
Lebanon, Uganda and other places in the world, we now
have government leaders who believe if we will sit down

and talk to and reason with these *jihadists* we will be able to live at peace with them.

There is a full-scale invasion across our southern border by some twelve million (and perhaps as many as twenty million) illegal immigrants that is changing the demographics of this land. If this invasion is not dealt with and if total amnesty is granted it will be the king of all train wrecks from which our nation will never recover.

Certain groups are working night and day to find a way to take our guns away from us. Remember that King George of England once tried that and it cost him his fat butt.

Social Security and Medicare are broke and racing toward bankruptcy. Those in charge of Medicare report they lose $50 billion a year but they don't even know where the money is going. The Congress (with only a twenty percent approval rating with the American people) and the White House have burdened us with unsustainable national debt and our government is literally drowning in red ink.

So millions of Americans now are asking: "If the federal government can't run Social Security or Medicare, why on earth would we want to trust them to run anything? Perhaps that is why a solid majority of Americans opposed the national healthcare law passed this year.

But there's more. We see the collapse of the U. S. dollar, thirty million unemployed or underemployed and Climategate where heretofore reputable scientists, through fraudulent scientific research, have used tricks to hide the truth and stifle other scientists who question both global warming and the fabricated hysteria of climate change.

Money seems to be burning a hole in the pockets of President Barack Obama and the men and women in Congress. They have committed trillions of dollars in stimulus plans, the bail out of Wall Street (that included $30 billion to prop up European banks), the General Motors and Chrysler bailout, and a costly national healthcare plan.

But the problem is that Congress is running out of *our* money!

I'm fearful that if the Obama administration continues this spending orgy, our nation will go belly-up and into bankruptcy with rampant inflation and unprecedented unemployment. Hopefully, the President will change course and stop the train wreck before it takes place. Otherwise, they will kill the Goose that laid the Golden Eggs and it will take generations to undo what is taking place in Washington today.

I also believe we face a national crisis of epic proportions, a spiritual malaise that is far more dangerous than even the threat of a nuclear holocaust. It is a crisis of the spirit, the soul, the conscience.

Historians will call this a fallen generation that lost its way and surrendered its existential will to the hostile and depraved where perversion and immorality were accepted but Christian values criticized as old-fashioned and bigoted. We have embraced things that would have horrified our parents' generation from unrestricted abortion-on-demand to extreme body piercing and teaching homosexuality to children in elementary schools.[1]

We are so enamored with the pursuit of pleasures and possessions we have lost our spiritual compass. The great ship of state is blowing here and there and everywhere and is about to sink but our elected officials in Washington, both liberals and conservatives, are standing on the sidelines like naked souls in a habitation of dragons eating the fruit of the poison tree.

Our nation is in peril and spinning out of control. Secular progressives, those we once called liberals, are determined to create a new world without any moral limits. One of these angry mornings the people of this land may wake up and realize we have lost our souls and are on the verge of national catastrophe as this feeble, hedonistic worldview spreads deeper and deeper into every area of

America's X-rated culture like an anvil dropped on the national conscience.

Things once universally accepted as criminal and rejected by decent people, now have, as Pope John Paul II said, "gradually become socially acceptable."

The feminists, the politicians and the news media have so conditioned us to disbelieve the truth that many of us have come to the place where we will accept a lie and believe that lie is all that makes sense.

Those of us who want to hold on to the good things of the past like values, free enterprise and the Golden Rule are called conservatives. But we are also called old-fashioned and intolerant by the secular progressives and their powerful allies in the news media. It has become commonplace for television news anchors and editorial writers to take a swing at anyone who stands up for traditional morality or speaks out against evil. However, the progressives who embrace every kind of immoral behavior and believe in government-driven change are the new "saints" in our land.

Could there be a time in the not-too-distant future when schoolchildren around the world will study the rise and fall of the United States and learn that it was just another great civilization that failed?[2] Will they study from a textbook entitled *Once Upon a Time in America?*

Our schools have been turned into indoctrination centers for social engineering where elementary-school students, in some areas, are given birth control pills without their parents' knowledge or consent. Secular architects are forcing homosexual education on the school kids in other schools and have turned our children into guinea pigs for their social experimentation.

Our land has become a sexual nuthouse as Hollywood producers and directors decided to break down most all moral barriers and created television programs laced with profanity and gratuitous sex. The music industry produced rap music that glorified rape, drugs, cop killing and satanic themes. The American people went to sleep and did nothing. Homosexuals/lesbians have emerged from the back street to Main Street and are working feverishly for same-sex marriages in numerous states. Such marriages have been declared legal in both Massachusetts and California and the movement is spreading all across the land.

Columnist Janet Folger in her book *Criminalization of Christianity* points out that there was only one other time in history when same-sex marriages were recognized – just before Noah's Flood. She found that information in both the Babylonian Talmud and in several Jewish traditions. Two Jewish rabbis have confirmed her research.

Should we take that as a warning?

Progressive futurists speak of a golden age where there will be no moral parameters of any kind and where people will be free to do whatever they please regardless of the pain and suffering it inflicts on others. They sugar coat immorality and paint a rosy picture of a hedonistic society that accepts every kind of perversion known to man.

The progressives are studying ways to get rid of the elderly, infirmed and helpless in our land because some argue they are a drain on our health care system. That's called "euthanasia" and it's on the progressives' agenda and not too far away.

Who would have believed that our standard of morality could have gone down so far so fast? Remember these headlines?

- Ten Commandments removed from Alabama Supreme Court Building.
- Prayer prohibited in public schools.
- Nativity Scenes banned from public property.
- Boy Scouts sued by homosexuals.
- Bible called "hate speech."[3]

The far-famed English historian Edward Gibbon who

wrote *The Decline and Fall of the Roman Empire* said that Rome fell because it rotted from within, lost its civic conscience and committed suicide as its citizens became immoral, soft and lazy. Gibbon noted that "the leaders of the empire gave in to the vices of strangers, morals collapsed, laws became oppressive, and the abuse of power made the nation vulnerable..."[4]

I believe that America is following in the tragic footsteps of Rome.

Although the Roman Empire is the classic example, there were other great civilizations such as the ancient Chinese dynasties and the Ottoman Empire that failed and were relegated to the dust bin of history.[5] Remember Carthage, the Greek city-states, the Holy Roman Empire, the kingdoms of France and Spain and the gradual decline of the British Empire?[6]

"They were historic catastrophes from which whole peoples did not recover for centuries," says Thomas Sowell, economist and author with the Hoover Institute at Stanford University. "It has been estimated that it was a thousand years before Europeans again achieved as high a standard of living as they had in the Roman times. The Dark Ages were called dark for a reason."

Here's something to think about: During the Eighteenth Century, a great religious revival swept across England fueled by the fiery preaching of John and Charles Wesley and George Whitefield. It was also an era of dramatic social and political change when slavery was abolished through the faithful efforts of William Wilberforce, a dedicated man of God.

During the Nineteenth Century, Charles Spurgeon preached in the Metropolitan Tabernacle in London and thousands came to hear him. Also, William Booth founded the Salvation Army during that period and ministered to the outcasts of English society in East End London – the alcoholics, criminals and prostitutes.

But today, only three percent of the English people attend church.

What happened? They died spiritually.

History has well-documented evidence that when a nation forgets it religious heritage, the nation dies.

Historian Will Durant noted, "There is no significant example in history, before our time, of a society successfully maintaining moral life without the aid of religion."[7]

Are we Americans walking down that same long and winding road?

Did the overt decline in morality in this land begin with Madalyn Murray O'Hare? Perhaps.

O'Hare filed a lawsuit against the Baltimore, Maryland, School District. Her petition said it was unconstitutional for her son William to be forced to pray in school and hear Bible readings. She also said that her son was persecuted by other students when he refused to participate in the religious exercise, a charge he later denied. The liberal Warren Supreme Court agreed with O'Hare and effectively banned prayer and Bible reading from public schools in the land.

Most Americans, unaware of the serious implications of the high court's decision, sat back and said very little and did nothing.

Ten years after the high court ruled against prayer and Bible reading, the justices approved abortion on demand and struck down all the states' law restricting the killing of the unborn. Although a few voices, primarily Catholics, cried out against it, most people said and did nothing.

The floodgates were open. Progressives in politics and the news media spread the damaging propaganda that whatever individuals do in private is their own business, even if it takes place in the Oval Office at the White House.

It's apparent a Sword of Damocles hangs precariously over the American conscience.

Lee Iacocca, the former president of the Chrysler Corporation, has written a book entitled *Where Have All the Leaders Gone?* In the book he deals with the geopolitical and economic cynicism and despair in the land.[7]

"Am I the only guy in this country who's fed up with what's happening?" he asks.

Iacocca says there are clueless leaders in Washington steering the ship of state right over a cliff and corporate gangsters stealing the people of America blind. "And we can't even clean up after a hurricane," he said.

"But instead of getting mad, everyone sits around and nods their heads when the politicians say, 'Stay the course.' 'Stay the course?' You've got to be kidding. This is America, not the Titanic."

Iacocca says he hardly recognizes America anymore and no one has the right to call himself a patriot if he or she is not outraged over what is happening in the land today.

So here's where we stand, according to Iacocca:

- The United States is engaged in a bloody war.
- We're losing manufacturing jobs to Asia.
- Our once-great companies are being destroyed by healthcare costs.
- Gas prices are out of sight and no one knows what to do about them.
- Our schools are in trouble.
- Our border with Mexico is like a sieve.
- The middle class is being squeezed by taxes and all the regulations coming out of Washington.

"These are times that cry out for leadership," he says.

The former Chrysler president says he isn't trying to be a voice of doom and gloom in his book.

"I'm trying to light a fire," he says. "I'm speaking out because I have hope – I believe in America… If I've learned one thing, it's this: You don't get anywhere by standing on the sidelines waiting for somebody else to take action…

That's the challenge I'm raising in this book. It's a call to 'Action' for people who, like me, believe in America. It's not too late but it's getting pretty close."

Chapter Two
The Death of Meaning

"We've reached a new day, when anything said, no matter how violently torn from context, is presented as however one wishes and accepted as fact by the interpreter. It's the death of meaning."

So says Bill Bennett, author, conservative political theorist and Secretary of Education during President Ronald Reagan's administration.

The death of meaning is synonymous with the rise and fall of absolute truth. But I'm not ready for the funeral.

There was a time when biblical imperatives such as the Ten Commandments, the Sermon on the Mount and the so-called Golden Rule represented a moral compass to guide us. They were standards of *absolute truth* and the exalted principles were believed and understood by most people.

Earlier generations had a clear understanding of absolute truth, right and wrong, morality and immorality and good and evil. They rightly considered the aforementioned biblical imperatives the guide stone for all areas of the American democratic experience: business, government, community, family, religion and a host of other bed-rock institutions. They believed that moral health was critical to the economic and governmental health in the land.

Then along came moral relativism that argues against absolute truth and excludes all moral imperatives and religious principles.

There is a distinct difference between moral relativists, i.e. secular progressives and traditionalists. Progressives believe that America is an intrinsically flawed nation and radical changes should be made. Traditionalists believe there are many good things about America that should be cherished forever.

I have also discovered that most of the secular progressives run in packs: those who favor abortions, evolutionists, homosexuals/lesbians, radical environmentalists, global warming alarmists, the feminists and most leaders of labor unions. Needless to say, these groups working together have a powerful voice in this land.

Blacks

The Honorable Clarence Thomas, Associate Justice of the United States Supreme Court, in a speech at Hillsdale College in Michigan, referred to the subject of moral relativism.

Thomas said that back in the 1980s, he read a book by Paul Johnson entitled *Modern Times*.

"One point it makes clearly is the connection between relativism, nihilism, and Nazism," Thomas said. "The common idea that you can do whatever you want to do, because truth and morality are relative, leads to the idea that if you are powerful enough you can kill people because of their race or faith."

Thomas also said: "So ask your relativist friends sometime: What is to keep me from getting a gang of people together and beating the hell out of you because I think you deserve to be beaten?"

"Too many people think that life and liberty are about their frivolous pleasures," he said. "There is more to life. And again, largely what relativism reflects is simply a lack of learning."

The humanist John Dewey, American philosopher and educational reformer, was one of those responsible for the premature death of meaning. He once said: "There is no

God and there is no soul. Hence, there are no needs for the props of traditional religion." Dewey, who lived from 1859 to 1952, was a co-author of the Humanist Manifesto I.

Dewey also proclaimed, "With dogma and creed excluded, then immutable truth is also dead and buried. There is no room for fixed, natural law or moral absolutes."[1]

Hence, the relativists were the ones who gave us the flawed dictum: "If it feels good, do it," regardless of any consequences to oneself or others. The implication is that the teachings of the Bible and Christianity are passé.

Within that framework, there are few moral taboos and young people themselves, not their parents or pastors of their churches, should decide what is right and wrong and how they should live their lives.

This teaching is the father of unwanted pregnancies, excessive drinking, cribbing on tests, sexually transmitted diseases and the recreational use of illegal drugs as traditional values have been abandoned under the guise of freedom and enlightenment. When truth is no longer relevant, it leads to the premature death of meaning.

The news media are the public relations arm of moral relativists. They are opposed to traditional Christianity, conservative talk radio and biblical moral values.

When broadcast or print journalists have a mindset on any issue, they routinely report it as true. They do not believe in absolutes unless it is something they *absolutely* believe.

For instance, they are convinced global warming is about to destroy this planet even though thousands of reputable scientists disagree with that proposition. I find it fascinating that the news media solemnly warns the world that the Arctic is getting warmer but fail to report that Antarctica is getting colder.

They have the same mindset on other scientific and cultural issues. For instance, these groups have accepted the flawed Kinsey Report as an absolute barometer of the sexual mores in this land. However, Dr. Judith Reisman has proven research that reveals Kinsey and his associates questioned many prison inmates, sex offenders, prostitutes and exhibitionists to come up with their conclusions regarding sexual practices in America.

Dr. Reisman says Kinsey's research also involved illegal sexual experimentation on several hundred young children.

She has written a book entitled *Kinsey: Crimes & Consequences* which describes his flawed research that has had such a profound influence on peoples' beliefs concerning human sexuality.

"Today, half a century later, Kinsey's unchallenged conclusions are taught at every level of education – from elementary school to college – and quoted in textbooks as undisputed truth..." Dr. Reisman says. "Yet Kinsey's grotesquely fraudulent research has served as the very foundation of modern 'sex science,' and his claim that one in 10 people are homosexual is central to the gay-rights movement." Some researchers believe gays make up from two to three percent of the population.

Although Kinsey has been discredited, secular progressives believe his conclusions are absolute.

I have asked myself the question: Why would secular progressives continue to use Kinsey's report and refer to it as a credible scientific/sociological text for sexual mores in this land? The answer is simple. The report recommends all kinds of sexual behavior that once were condemned as immoral. That fits in well with the progressives in the media and academia who want to rid our land of all sexual taboos.

The progressives also accept evolution and abortion as unquestionable absolutes, but do not question sex outside of marriage, homosexuality or pornography which they do not consider moral dilemmas.

Remember, the progressives must have issues or their movement would die a slow death. Hence, subjects such as global warming are agenda-driven like the Kinsey report.

They consider the concept of the "separation between church and state" absolute dogma although the Constitution has nothing to say on the matter.

Thomas Jefferson used the term in response to the Danbury Baptist Association of Connecticut in 1802 when they raised the question of religious freedom.

However, progressives accept the term as synonymous with the First Amendment to the Constitution.

Hence, we have even seen the death of the traditional meaning of the Constitution.

Dr. Donald Wildmon, a Methodist minister and founder and president of the American Family Association, says that the secular progressives have been so successful in spreading their worldview that we in this land now are facing "America's Moral Katrina."

He gives the following examples to support his thesis:

- Scenes and language on television are much more explicit than ever before. "If the rate of decline continues, it will not be long until we see hard-core pornography on the networks."

- Big-screen movies also have become "raunchier" with profanity, gratuitous sex, and blasphemy against God. "Restraint and reason are passé."

- Popular music is becoming much more explicit and he says studies reveal that listening to this music can lead young people to abandon their parents' values.

- Alcohol and drug usage among young people is growing as the users are getting younger and younger.

- Abortions are now accepted as a fact of life by a large segment of our society.

- The bias in the news media against conservatives and Christians is getting stronger every day. Fox News is one of the only exceptions.
- Nearly every day the people of this land hear of the frantic efforts to legalize homosexual "marriage." Ford Motor Company, Wal-Mart, McDonalds and dozens of other major corporations have given their approval and money to support the homosexual agenda.

"Meanwhile, most Americans go on with their day-to-day-lives raising children, meeting family needs, working, etc.," Wildmon said. "Millions sit in their houses of worship on Sunday morning, and see no connection with the moral decay around them."

He warns that the secular progressives will never be satisfied "until all influence of the Christian faith is removed from the public arena."

"Our religious freedom, which helped make this country the greatest ever known, is being stripped away," he said.

With the death of meaning, have we also witnessed the death of sin?

Psychiatrist Dr. Karl Menninger in 1973 asked that question in his book *Whatever Became of Sin?* He was a member of the famous Menninger family and one of the founders of the psychiatric clinic in Topeka, Kansas, that bears the family name.

Menninger was concerned that the people of this land no longer follow a divine standard of right and wrong.[2]

"The word 'sin,' Menninger maintained, is disappearing from the American vocabulary," said Kelly Boggs, writing in *The Christian Index*. "The definition of sin the... doctor had in mind was not sociological... No, Menninger's denotation of sin was religious and included the idea of willful rebellion against the standards of God."

Menninger believed that simply removing the word "sin" from our vocabularies would not make it go away.

"Menninger was correct in his analysis," Boggs said. "In the three decades since his views on sin were first published, the word in its religious sense has slowly disappeared from use in American society. As a result, the idea of transgressing a divine standard has become nothing more than a concept rooted in nostalgia."

I wonder what Menninger would think about our present crisis of morality in America. He certainly would repeat the question: Whatever became of sin? He might also ask: Whatever became of shame?

"Like any good doctor, Menninger prescribed a solution to the problem of 'vanishing' sin," Boggs said. "Menninger called on America's clergy to: 'Preach! Tell it like it is. Say it from the pulpit. Cry it from the housetops.'"

USA Today's Kathy Lynn Grossman asked the question, "Is sin dead?" She approached the subject in her article for the paper titled "Has the 'Notion of Sin' Been Lost?"

Grossman's article raises the issue of what people, including preachers, believe about sin.

"She also points to a question that should trouble the Christian conscience: 'How can Christians celebrate Jesus' atonement for their sins and the promise of eternal life in his resurrection if they don't recognize themselves as sinners?'" Dr. R. Albert Mohler, Jr., president of the Southern Baptist Theological Seminary in Louisville, Kentucky, asks. "That demands an answer... We are reminded yet again that an understanding of sin is preliminary to understanding the Gospel... The magnitude of our sin explains the necessary magnitude of Christ's atonement."

We have also lost the meaning of words and terms that once highlighted beliefs based on substantive principles but now are disguised in ambiguity.

For instance, during the frenetic 2008 presidential campaign, then-Senator Barack Obama shouted, "Yes, we can!"

Any thinking person would respond, "Can what?"[3] The obvious answer is "Hope."

"Hope for what?" asks Charles R. Kesler, professor of government at Claremont McKenna College.

Senator Obama then would respond, "For change" or "Change We Can Believe In," as opposed to "Unbelievable changes."[4]

"But the elementary problem with this, which any student of logic might raise, is that change can be for the better or for the worse," Kesler says.

According to Kesler, secular progressives believe that change is synonymous with improvement.

"They fail to weigh the costs and benefits of change, to consider its unintended consequences, or to worry about what we need to *conserve* and how we might go about doing that faithfully," Kesler says. "They ask Americans to embrace change for its own sake, in the faith that history is governed by a law of progress, which guarantees that change is almost always an improvement. The ability to bring about historical change, then, becomes the highest mark of the liberal leader."

Kesler points out that the secular progressives want more government while conservatives want less. But he says there has been a falling away from the standards of Ronald Reagan's conservatism and it was replaced by the "compassionate conservatism" of George W. Bush which, in essence, has led to more government, very little different from the secular progressives.

A giant first step toward the resurrection and rebirth of meaning would be for the preachers in the pulpits all across America to boldly proclaim the absolute truth of God's Word to the millions of Christians in this land. That, in turn, would empower believers everywhere to "always be ready to give a defense to everyone" for the hope that is in us as we are encouraged by the Apostle Peter in his first epistle chapter three and verse fifteen.

Chapter Three
Whatever Became of America?

I often lament the fact that the land of my childhood is gone forever. It was a wonderful land where people looked out for one another and helped them survive the hard times of the Great Depression, ministered to those who were sick and mourned the dead. My father Claude Keith ran his own welfare program. He would place bushel baskets of fruit from our orchard out in front of the house for the impoverished Cherokee Indians of Northeastern Oklahoma so they wouldn't have to go hungry or beg for food. My mother Hattie Mae Keith also helped out by feeding the hungry hobos who came through our town of Jay, Oklahoma, on the train. She always gave the hobos the same food she had prepared for the family.

There was virtually no crime, particularly in the rural areas. But every few years there would be a killing and it was the talk of the town for months since it was so unusual.

Back in the 50s, only a handful of the more daring kids in our high school drank beer and a few smoked cigarettes. There were no drugs or unwanted pregnancies.

We did some crazy things like writing "Kilroy Was Here" on the sides of buildings. I never learned who Kilroy was. But it was fun. We also soaped neighbors' car windows on Halloween night and turned over outdoor toilets, occasionally with someone inside.

That behavior was endemic to our culture when right and wrong were as simple as black and white. But this country has changed so drastically in the past fifty years I hardly recognize it anymore.

Today we live in a culture that has done a one-eighty toward bizarre behavior now so deeply entrenched some think it's normal. We are witnessing a cultural catastrophe spinning at warp speed through the early days of the Twenty First Century and no one knows for sure where it is taking us.

Let's take a closer look at where we are today and see if we can make any sense out of a lot of nonsense in our culture and the effect the cultural shift has had on the people of America.

We saw the culture begin its downward spiral after World War II. It hit bottom in the 60s and 70s.

John Lennon, one of the Beatles, struck a chord in the lives of the younger generation in "Imagine," his utopia-themed song written during the height of the Vietnam War. I'm sure most of you remember the words: *Imagine there's no countries; It isn't hard to do. Nothing to kill or die for; And no religion too. Imagine all the people, living life in peace.*

Lennon acknowledged the song was anti-religious and anti-American and said it was the perfect Communist Manifesto. *Rolling Stone* magazine, which is not my favorite source of news and commentary, says that "Imagine" is one of the top three songs of all time.

Melanie Morgan wrote a column entitled "Imagine No Hippies." Speaking of Lennon's "Imagine," she said, "The words of that song shoot up my blood pressure... We all want to live in peace, but no country? No religion?"

Lennon's song became the battle cry for thousands in the hippie and anti-Vietnam War generations of the 1960s and set the stage for the radical changes that led our culture into a moral decline of epic proportions.

That causes me to wonder: How could we have fallen so far, so fast?

The pursuit of pleasure is a dominant driving force in

American life today.

Mohler the theologian warns us that the people of this land are saturated with sexual immorality. He expressed that view on CNN's special "God, Sex and Greed." The program focused on America's simultaneous fascination with religion, lust and the pursuit of possessions.

"It's a sign of a culture that is increasingly seeking gratification in all the wrong places," Mohler said.[1]

He said that explicit immorality once was virtually hidden in dirty books and magazines and smutty movie houses but now we see it on billboards, in shopping malls and advertised on T-shirts that young people are wearing. The American people have absorbed this immorality into the heart and soul of our culture.

Every day we see more and more evidence of a culture gone bad and wonder where are the other voices who, like Mohler, would cry out against the causes of our moral decline and call us away from obscenity back to sanity?

Keith Phillips, founder and director of an inner-city organization known as World Impact, presents clear evidence of the disastrous moral decline: Seventy-five percent of all children born in big city hospitals will be born to single mothers; fifty percent of these children will have some kind of drug addiction; there are more young black men in prison than in our colleges and universities; and in Los Angeles County alone, there are ninety-thousand homeless and some forty percent of those are women and children.[2]

All of us were shocked when we learned that Michael Vick, the star quarterback for the Atlanta Falcons, was caught in illegal dog fighting.

Michael was the poster boy for the great American success story for he grew up in a poor family in the projects of Newport News, Virginia. The Falcons gave him an unprecedented one hundred-thirty-million-dollar contract and he made more millions in endorsements for Nike, Coca Cola,

Kraft and the Rawlings sports company.

A federal judge sentenced him to twenty-three months in prison and three years' probation for his part in the dog-fighting conspiracy.

His behavior may have cost him as much as ninety million dollars.

Through this very difficult experience, Michael has experienced a personal transformation and today is the star quarterback for the Philadelphia Eagles and I for one am glad to see the transformation that has taken place in his life.

I was amazed that there were so many voices calling for the judge to throw the book at Michael for his cruelty to the animals.

Now understand, I love dogs and it grieves me to see them, or any other animals, mistreated.

But there was something quite hypocritical about the way our culture condemned Michael Vick. Yes, he killed some dogs and others died fighting for their lives and concerned animal activists were right when they cried out against the cruelty.

However, where are the voices today crying out for the four thousand unborn who are killed in their mothers' wombs every day in this land? That's the Hypocrite Factor, the ring around the collar that secular progressives don't want to discuss. But it reveals a culture in crisis.

Picture this: wealthy doctors who live in multi-million-dollar homes, drive Mercedes Benz automobiles and have offices in the fashionable areas of our cities kill the unborn with impunity while Michael Vick goes to jail for dog fighting. Yet the doctors are respected members of country clubs and pillars in the community and some even attend church.

Tell me, what's wrong with that picture?

Drinking has replaced baseball as the great American

pastime. Binge drinking – that often leads to death – is another societal indicator in this culture. Beer- and whiskey-fueled beach brawls are common among students during spring break on the beaches of Florida, California, Cancun, Mexico, and Padre Island in South Texas.

So many young movie stars and sports heroes have been picked up for DUIs, their theme song should be "Whiskey at the Wheel."

Police have arrested a number of professional football players for drunken driving. Some teams have more arrests than wins.

There was a brawl at Pacific Beach in San Diego during a Labor Day weekend and sixteen people were injured. Police officers had to wear riot gear and use smoke and tear gas to break up the alcohol-fueled melee when the hostile crowd of about five hundred threw bottles and cans at the officers.[3]

Let's take a look at what I call "the Hypocrite Factor."

I've often wondered why there would be such stringent controls on cigarettes and other tobacco and virtually no controls on alcoholic beverages.

Smoking is dangerous because it hurts people but I believe there has been a lot of knee-jerk discrimination against smokers. For instance, the price of a pack of cigarettes is out of sight as is a can of Skoal or Copenhagen, favorites of the "good old boys never meanin' no harm" who like to dip now and then. Also, tobacco manufacturers are denied advertising on TV and print ads must carry a warning from the Surgeon General.

What about smoking pot?

Medical marijuana is now available in California. All you have to do is drop into a dispensary for a fix. You do need a note from your doctor. You can get the free joint for just about any ailment including inner-grown toenails and dyspepsia. If you are a woman and high heels hurt your feet or if you have a skin rash, get a fix.

San Francisco Mayor Gavin Newsome plans to put a tax on Cokes, Pepsis and other soft drinks to fund the twenty-seven medical marijuana centers in his city where you can get stoned out of your head for a week to relieve the tooth ache.

But what about drinking?

Have you watched any sporting events lately? Beer commercials are everywhere. Actually, sports and beer drinking are almost synonymous.

There's no warning from the Surgeon General that drinking can be harmful to one's health. There's no oppressive taxation on liquor and beer. Perhaps there should be a warning like this on every fifth of whiskey and can of beer: "The Surgeon General has issued a warning to all young ladies that eighty-five percent of those who get pregnant before marriage do so under the influence of some form of alcoholic beverage."

Although it would really stretch the imagination, the Surgeon General could add, "And many of those pregnancies will be terminated by a doctor before the baby ever sees the light of day."

I know that is only wishful thinking.

There are very few present-day voices warning people about drinking for it is so much a part of our culture. Hence, most pulpits in our churches are silent on the issue. One church has sanctioned a so-called "Beer and Bible Club" that meets in a local bar and the serving of beer and wine at a home church meeting.

But there is a voice out of the past that we should hear, the voice of Evangeline Booth, daughter of William Booth, the founder of the Salvation Army.

Drink has drained more blood,
Hung more crepe,
Sold more houses,
Plunged more people into bankruptcy,

Armed more villains,
Slain more children,
Snapped more wedding rings,
Defiled more innocence,
Blinded more eyes,
Twisted more limbs,
Dethroned more reason,
Wrecked more manhood,
Dishonored more womanhood,
Broken more hearts,
Blasted more lives,
Driven more to suicide, and
Dug more graves than any other
Poisoned scourge that ever
Swept its death-dealing waves
across the world.

The Eighteenth Amendment to the Constitution banned the manufacture and sale of all alcoholic beverages in this land from 1920 to 1933. That era was commonly known as "Prohibition."

Republican Christians in the north and democrat Christians in the south joined together to pass the amendment.

Critics say it was a time of black market booze and racketeering as underworld figures such as Al Capone of Chicago and others made millions on the illegal traffic of whiskey and beer.

During Franklin Roosevelt's campaign for the presidency in 1931, he promised to repeal Prohibition and he kept his promise. Today, the social progressives in our land ridicule Prohibition as though it was something evil.

There was nothing wrong with Prohibition. If it had not been repealed, millions of people would have been much better off and hundreds of thousands of others would be alive today.

Do you know what the problem was?

Jerry Falwell once said, "When Christians win, they quit, and when they lose, they quit."

The Christians in this land went to sleep. They demanded no accountability from crooked judges and high-ranking police officers who took payoffs from the mobsters and allowed them to sell booze. The people should have voted them out of office but did nothing. And while the Christians slept, this country went through one of the most corrupt periods in its history. And Prohibition took the blame.

Parenthetically, deaths on the highways are the number one cause of death among young people and most of those involve drinking.

I believe it's evident that our culture is facing a moral collapse and we see it clearly in the relationship between morality and suicide, particularly among young people.

Suicide is the third leading cause of death among young people in this land.

Now a UCLA psychiatrist says promiscuity, which she believes is the root cause of much depression, often leads to suicide.[4] She adds that the promiscuity factor is being ignored by most doctors who treat patients for depression.

"There are approximately 1100... suicides on our campuses every year – a very, very startling number," says Dr. Miriam Grossman, a psychiatrist at UCLA.

Dr. Grossman believes that depression is at epidemic levels on college campuses. Those in the medical community treat various symptoms of depression such as anxiety, insomnia, eating disorders and other emotional problems but do not deal with the suicide/morality connection.

There is an "anything goes sexual free-for-all" that exists on college campuses, she points out.

"Parents need to know that dating is pretty much dead these days," Grossman says. "College students hang out in groups, and then they pair off and 'hook up.'" She explains that a "hook up" is an unplanned sexual encounter between two students who may never see one another again.

"So you can understand that it's a no-strings-attached, no-emotions-attached, casual sort of set-up," she explains. "And this is different than the way it was in past generations, because it's normalized... and we have between 40 and 80 percent of college students participating in 'hooking up.'"

She warns that there are rampant physical symptoms of depression and certain biological changes that take place after sexual contact and there are emotional, physical and spiritual after-effects of immorality.

Many of the UCLA students who come to her for help have followed the guidelines for "safer sex" but eventually wind up in her office, casualties of the "hook-up" culture.

Many American young people worship at the altar of a cheap orgasm and believe that a few minutes of physical ecstasy is the pathway to reality.

Profanity is another sign of the cultural times.

Several years ago news icon Paul Harvey lamented the terrible use of profanity during professional basketball games. He said most everyone in the great arenas could hear the players' foul language, including children and teenagers there to watch the game.

I have wondered why the owners in general and the coaches in particular have not stopped the profanity. Is it because through the years they have become so accustomed to foul language they no longer believe anyone cares?

As Gordie Howe, the great professional hockey player once noted: American sports fans are bilingual – they speak English and profanity, according to Kelly Boggs writing for Baptist Press.

Now foul language has filtered down to the students of our colleges and universities who turn the air blue cursing the referees and opposing teams and their fans.

According to a report in *USA Today*, several universities have cracked down on profanity during sporting events.

Boston University outlawed profanity from its hockey arena in 2006. When some of the fans refused to stop cursing, the university suspended their ticket privileges.[5] Also, the University of Florida adopted a code of conduct for its fans and ejected several from all the school's sporting events.

"If you have been anywhere near a football sideline recently, you likely heard language that would make the devil proud," Boggs said.

Boggs asks the question: "How bad is the language among coaches and players in football?" It is so bad that both head coaches Tony Dungy formerly with the Indianapolis Colts and Lovie Smith of the Chicago Bears made headlines before Super Bowl Forty-One in Miami, Florida, because neither uses foul language.

Isn't it amazing that the non-usage of profanity became national news?

But foul language is not confined to the large sports palaces of this land. You can hear it when walking through a mall or when you turn on the television.

I was deeply disturbed when I heard that Dana Jacobson, an anchorwoman with ESPN, had used vile, blasphemous language as she trashed the name of Jesus at an event to honor two of her colleagues. Her irreverent language was an insult to Christians everywhere.

Remember Don Imus? He was fired when he made disparaging remarks against a group of fine young black women athletes. And firing him was the right response to his improper slur of the young women.

Imus has never been on my A list and I've never listened to any of his programs.

But once again there is a Hypocrite Factor when you compare the remarks of Don Imus to those of Dana Jacobson.

I believe that if Jacobson had said what she did about blacks or homosexuals, ESPN would have fired her. But she cursed Jesus who isn't here to defend himself, to organize a march, to whip up support from ambulance chasers like Al Sharpton and Jesse Jackson, the NAACP, the ACLU and the People for the American Way.

Concerning Jacobson, ESPN probably knew that Christians would wimp out and say nothing in defense of our Lord for, if the past is the pattern, most Christians are sound asleep and either don't care or aren't aware a culture war is being waged for the soul of America.

I protested to ESPN and received a generic reply that Jacobson had been suspended for a week and that she is sorry that she offended anyone with the remarks she made about Jesus.

I don't intend to ever watch Jacobson's program again.

ABC, the parent company to ESPN, fired actor Isaiah Washington, star of *Grey's Anatomy*, for an anti-gay reference to fellow actor T. R. Knight. But when Jacobson made the blasphemous remarks about Jesus, ABC gave her what amounted to a week's vacation.

"ABC has two standards, one for anti-gay comments and one for anti-Christian comments," said Donald Wildmon of the American Family Association. "Those who use anti-gay comments are punished. Those who use anti-Christian comments are supported."

Another indicator of culture in decline is how we dress.

"Overexposed" is the word that comes to mind for many Americans when we consider young celebrities such as Paris Hilton, Britney Spears, Lindsay Lohan and Nicole Ritchie, according to Penna Dexter, a columnist for Baptist Press.[6]

Dexter believes we see too much coverage of them in the news and not enough covering on their bodies.

"But it's not just Hollywood," Dexter says. "Young women in general are looking a bit trashy these days."

Dexter says that even a lot of good girls dress like they are bad.

"Sometimes even their mothers, unwilling to look matronly, find themselves with scant middle ground between frumpy and 'Desperate Housewives,'" she says. "So they compromise, just a little, then a little more until we have a new norm where fashion trumps modesty."

Dexter believes the fashion industry may have conspired with the popular dress culture to tear down natural modesty among girls and young ladies. "It's almost impossible to find clothes teen girls like that don't reveal too much, sometimes way too much."

Theologian Warren Wiersbe gives us some insight into our present-day moral dilemma: "Perhaps our problem today is what C. S. Lewis pointed out: we don't hate sin enough to get upset at the wickedness and godlessness around us. Bombarded as we are by so much media evil and violence, we've gotten accustomed to the darkness." [7]

Chapter Four
Big Brother and the
Fate of a Nation

On each landing, opposite the lift shaft, the poster with the enormous face gazed from the wall. It was one of those pictures which are so contrived that the eyes follow you about when you move. BIG BROTHER IS WATCHING YOU, the caption beneath it ran.

"Big Brother Is Watching you" from George Orwell's novel *1984* was prophetic and we see the prophecy being fulfilled today in every area of American life.

The novel clearly exposed the dangers of powerful, intrusive government in the lives of people and equated Big Brother to any omnipresent, omnipotent and intrusive government that governs people from the cradle to the grave.

Norman Thomas, this land's premiere socialist thinker and leader, predicted that Big Brother would one day take absolute control of life in this land. He said: "The American people will never knowingly adopt socialism, but under the name of liberalism they will adopt every fragment of the socialist program until one day America will be a socialist nation without ever knowing how it happened."[1]

Thomas Jefferson warned us about that kind of government. He said, in essence, government that is big enough to give you everything you want is big enough to take everything you have.

President Franklin Roosevelt took the first giant steps

toward socialism, in the name of liberalism, back in the 1930s and 1940s. He instituted various reforms to end the Great Depression but greatly exceeded all constitutional authority by creating certain unnecessary federal assistance programs, many of which we still have today.

Later, in the 1960s, President Lyndon Johnson expanded the welfare state in the land in an attempt to wipe out poverty. The liberal president threw some five trillion dollars at poverty through various welfare programs and wealth redistribution under the banner of the Great Society. But the programs failed.

Now Joseph Farah, founder of the WorldNetDaily, believes that Norman Thomas was right about America's swing toward socialism as Big Brother reaches into every area of American life.

Farah believes national health care is "a power grab by Washington in which the federal government will seize full control of another one-seventh of the U. S. economy."

He points out that the architects of the healthcare takeover hide the command-and-control nature of the new plan from the public.

"They are not asked by the press to show the American people even one successful program government has run," Farah said. "Yet the American people seem ready to put the lives of their children and grandchildren in the hands of Washington bureaucrats."

Why are the American people so opposed to nationalized health care? We've all heard the horror stories from other countries that have socialized medicine. Stories about long lines, incompetent treatment and waiting lists. Can you imagine a bureaucrat telling you if and when you can have a surgical procedure, who will be your surgeon and the hospital where it would be performed?

Canada has such a program for its citizens. There are news reports that mothers in British Columbia are experiencing a baby boom but are coming to the United States in droves to have their babies. And there are indications that the Canadians are fed up with their government healthcare system.[2]

According to a report in Fox News, some forty mothers were airlifted to the states because Canadian hospitals didn't have room for the newborns.

The health care program in Canada is in trouble as officials admit there is not enough tax money to purchase the equipment to handle high-risk births.

"The Canadian healthcare system has used the United States as a safety net for years," said Michael Turner of the Cato Institute. "In fact, overall about one out of every seven Canadian physicians sends someone to the United States every year for treatment."

In his latest slick-flick "Sicko," Michael Moore praised the United Kingdom's National Health Care Service. But a record number of Brits must have missed Moore's movie for they are going abroad in droves seeking better care, WorldNetDaily reported.

According to the *London Sunday Telegraph,* some seven thousand will travel to various parts of the world for treatment and that number was expected to increase to two hundred thousand a year in 2010.

Why are the Brits turning their back on their government health care program? Apparently they are fed up with having to wait weeks and months for treatment, the sub-standard quality of the treatment and the growing hospital-infection crisis, according to WorldNetDaily.

"How we could even be debating ideas like this (nationalized health care) in the 21st Century, after all of the climactic failures of socialism around the world, is amazing to me..." Farah said. "Americans may simply be too far gone spiritually, morally and intellectually to reject the temptations of socialism."

Farah says the promise of socialism is "one of the great lies of all time...You can have it all right here on Earth. You can live in utopia..."

"That's the essence of socialism," he said. "And it is finally seducing America as it has seduced much of the rest of the world over the last century."

Dr. Lyle Rossiter, a psychiatrist and author of the book "The Liberal Mind: The Psychological Causes of Political Madness," says: "Based on strikingly irrational beliefs and emotions, modern liberals relentlessly undermine the most important principles on which our freedoms were founded."[3]

"Like spoiled, angry children, they rebel against the normal responsibilities of adulthood and demand that a parental government meet their needs from cradle to grave," he said.

Dr. Rossiter says that any social scientist that has a basic understanding of human nature must not dismiss the importance of "free choice, voluntary cooperation and moral integrity – as liberals do."

According to Dr. Rossiter, the liberal/socialist/Big Brother agenda preys on human weakness and feelings of inferiority by "creating and reinforcing perceptions of victimization; satisfying infantile claims to entitlement, indulgence and compensation; augmenting primitive feelings of envy; and rejecting the sovereignty of the individual, subordinating him to the will of the government."[4]

New York Times columnist Roger Cohen wrote a column in December of 2007 titled "Secular Europe's Merits." Carefully and in great detail, he outlined why he prefers the liberalism/socialism of Europe to the more religion-oriented America.

Cohen asserts that the culture wars in America have led to "the injection of religion into politics..."

However, columnist Dennis Prager raises the question: Are those who have values based on religious convictions to stay out of the political process? Are only those who rely on secular sources allowed to voice their opinions?

He points out that Martin Luther King often used religion to fight against racial prejudice. Yet the liberal/socialist media establishment never criticized King for injecting religion into politics.

"The leftist argument against religious Americans' 'injection of religion into politics' is merely its way of trying to keep only the secular and religious left in the political arena – and the religious right, primarily evangelical Christians, out," Prager said.

Prager says the secular and religious left has embraced "myriad irrational hysterias" in his lifetime. For instance, the liberals/socialists believe the flawed research that unless there are extreme measures of population control, the people on planet earth will starve to death; that large numbers are dying from heterosexual AIDS and secondhand smoke; and, finally, humanity faces apocalypse because man is about to destroy the earth through global warming.

"It is extremely revealing that with regard to global warming scenarios of man-induced doom, the world's most powerful religious figure, Pope Benedict XVI, has just warned against accepting political dogma in the guise of science," Prager said.

Have you heard about Big Brother's plan to silence talk radio?

Talk radio, with hosts like Rush Limbaugh, Sean Hannity, Bill O'Reilly, Laura Ingraham, G. Gordon Liddy and others, is under attack and liberals in Congress are making plans to silence them forever.

They have been thorns in the side of Big Brother's liberal/socialist establishment in Washington for more than two decades. But when those hosts led a rebellion to stop Congress from giving amnesty to twelve million illegal immigrants – and millions of Americans responded as the plan went up in flames – the congressional power brokers decided something had to be done.

Whistleblower Magazine in the August, 2007 edition, said that the so-called "Old Media" – television networks, major daily newspapers, news magazines and the Associated Press – represent a worldview that is to the left of most of the people of Middle America. However, the "New Media" – including talk radio programs and the Internet – challenge the leftist media monopoly and provide balance and fairness.[5]

So now Big Brother is concerned and the sharp knives have come out.

"Some of the attacks on talk radio are quiet and technical – like proposals to force minority ownership on radio stations in an effort to get a more left-leaning worldview on the air..." Whistleblower pointed out. "Other attacks are more public and sensational – like the efforts of Jesse Jackson and Al Sharpton against talk show hosts."

After successfully leading the charge against talk-show host Don Imus for his inflammatory racial statements about a group of young women athletes, Sharpton said, "It is our feeling that this is only the beginning. We must have a broad discussion on what is permitted and not permitted in terms of the airwaves."

Senator Diane Feinstein of California says she may reintroduce the Fairness Doctrine in congress. That would require any talk show host to balance his/her views with opposing views. That, in essence, would kill talk radio.

She told Chris Wallace of Fox News: "Well, in my view talk radio tends to be one-sided... It' explosive. It pushes people to...extreme views..."

But what about the anchors and reporters for the national networks? Their telecasts often reveal a distinctively leftist worldview. Senator Feinstein never mentioned that they would be required to give equal time to opposing views and to me that is the epitome of hypocrisy.

The emergence of conservative talk radio is one of the great success stories in the history of broadcasting as millions of listeners each day tune in to hear the hosts and the talk-back from Americans as they discuss the burning issues of our time.

Why is talk radio so successful?

I believe there are people all throughout this land who are fed up with the leftist swing of television network news and are hungry for a more open and balanced discussion of current events in the land.

Chapter Five
Political Paralysis

As I was walking up a stair
I met a man who wasn't there.
He wasn't there again today.
I wish, I wish he'd stay away.
~Hughes Mearns

Big salaries, free travel throughout the world, a two-day work week with limited legislation but unlimited generosity to themselves and their families – that's our Congress!

They receive $174,000 a year, raise their own salaries and have free postal service worth thousands more to each of them and are not included in the new healthcare law passed in early 2010. The congressmen and women and their aides have a private plan paid for by the taxpayers.

They travel the world – ostensibly on fact-finding missions – in private or government jets and then roam their districts shoring up their voter base and telling the people all the good things they have been doing for them in Washington.

When I hear them talking about how hard they work, I want to say, "Strike up the violins!"

"We the People" has become a big joke to those in Washington who believe it should read, "We the Politicians." What we have in this land is freedom for our government and the tyranny of taxation for the people. Every day a little more of our freedom is being stolen from us by those in Congress who say they want to take care of us.

Ronald Reagan once said the nine most terrifying words in the English language are, "I'm from the government, I'm here to help you."

Simple solutions are *anathema* to the men and women in Congress who have a penchant for exaggeration and confusion on just about every issue that comes before them.

There is supposed to be a remedy for oppressive, intrusive government. It's known as the Tenth Amendment whereby sovereign states are aligned with, but not controlled by, the federal government. I believe this is what the original framers of the Constitution had in mind when they debated the delegation of authority, some to the government, but most to the states. Now it's turned upside down and Washington controls nearly everything we do and makes us pay a terrible price through taxation and hundreds of laws created to control our lives.

The tragedy is not that we have congressmen who are indifferent to the will of the American people but that the American people are asleep. We relinquish the mandate of "We the People" to the politicians and they continue to get reelected by people who refuse to write a letter, send an email, make a phone call or vote to get rid of the politician who acts like he/she is a member of the ruling class rather than the serving class.

That makes me feel like I've been kicked right in my cocoanut!

A seat in congress is not a position of privilege! And we should remember the congressmen and women are appointed and not anointed.

However, some well-meaning people will say, "No one is listening, and our complaints just give the politicians a headache, nothing more."

It is true many of them have contempt for the will of the people but if enough of us voice our opinions, they will listen. Thousands, even millions of us must not let them forget that they are working for us.

Sometimes I want to laugh or cry when I hear politicians talking about "our money." They don't have a dime. What they have is the trillions of dollars that the working people of America send them each year. On a scale of one to ten – one being "throw the bums out" and ten "enshrine them on Mount Rushmore," – I would give Congress about a one, maybe a two, no more. And most American people seem to agree with me.

In the Spring of 2010, according to several national polls, only twenty percent of the American people trusted Congress to solve the serious problems we face in this land. That means that eighty percent of us don't believe the congress is doing a good job.

So what are we going to do about it?

Our politicians tell us a lot about ourselves – the revelation that we have lost the passion for good government and that many of us are too lazy even to hold our elected officials accountable. We have kept silent too long and silence is consent. Where did we lose the wonder of it all? Why have we not challenged a government that wants to disregard the will of the people? Freedom was given to us by God. We only lend it to government.

Where are the prophets that would challenge the politicians' flatulent rhetoric?

The people never win. Politicians win and even, at times, criminals. But the people are pushed around by high taxes, excessive government control, congressmen who forget it was the people who sent them to Washington, and the federal courts that rewrite the Constitution to fit the judges' personal agendas.

"They have not been able to pass important legislation on…immigration reform or anything else of importance," said Dick Morris and Eileen McGann in their classic book *Do-Nothing Congress – Big Salary, Little Work, Free Trips.* "Instead, they spend their time raising money for themselves, bickering and passing bills to change the names of courthouses and post offices, commending winning sports teams, and suggesting that the flag be flown on Father's Day."[1]

Each year they are in recess a total of sixteen weeks.

However, Morris and McGann point out Congress was in session only 103 days in 2006. That's a little more than two days a week. That's good work if you can get it.

"Members of Congress have the only job in the country whose occupants can set their own salary without regard to performance, profit, or economic climate," said Tom Schatz, president of the Council for Citizens Against Government Waste. "Clearly, members must think that money grows on trees."

Each year they vote themselves another raise of several thousand dollars.

Schatz added that they are out of touch with reality when they continue to raise their own salaries.

"They forget that their salaries are paid by taxpayers," he said. "Americans are being forced to tighten their belts – if they even have a job."

"In the current Congress... the House schedule is laughable..." Morris and McGann said.

They even took one day off to watch a football game.

"And when they are technically in session, they don't do much," Morris and McGann said. "Take the month of February (2007), for example: the House was only in session for nine days – and on three of those days, the sessions lasted less than 20 minutes, while a fourth lasted for 39 minutes."

That same month, sixty-six members of the House took vacations at the expense of lobbyists, many to exotic countries.[2]

"Fourteen members and their spouses spent five days... at a luxurious hotel in sunny San Juan, Puerto Rico at an Aspen Institute conference on 'No Child Left Behind,'" Morris and McGann said. "Several of the members apparently took the conference mandate quite literally and brought their own children for a free trip..."

House Speaker Nancy Pelosi suggested that the House members' adult children be allowed to accompany members on trips paid by taxpayers.

House members traveled at government expense to the following cities and countries: San Juan, Vancouver, Prague, Grand Cayman, Florence, Helsinki, Punta Mita (Mexico), China, Barcelona, Montega Bay, Jamaica, Rome, Moscow, Cancun, Venice, Dublin, Istanbul, Honolulu and Krakow.

Morris and McGann ask: "What's wrong with free travel?"

"Well, for one, it creates a sense of entitlement," they say. "Members of Congress have gotten used to being wined, dined and flown to beautiful and expensive places. It adds to the insulation from their constituents, it takes up time that should be spent on the job they were elected to do and it basically provides tax free income for free travel."

Did you know that congressmen and women are allowed to lease SUVs and other expensive cars at taxpayer expense? And guess who pays for the gas, insurance and upkeep on these cars? The taxpayers.

"Congressman Charles Rangel was recently seen getting out of his Cadillac DeVille, which he leases for $774 per month..." CBS News reported. "And how about this one: Congressman Gregory Meeks was recently seen waiting for Congressman John Conyers to step out of Meek's Lexus LS460, which Meeks leases for $998 a month."[3]

However, CBS reported that members of the U. S. Senate are not allowed to lease cars to be paid for with public funds.

I believe it's time for congressmen and women to go back to work and make the tough decisions on the serious issues facing our nation.

It angers me when I realize that our Congress is responsible for the high gas prices we are paying at the pump.

"It's time for consumers to strike back against the real culprits behind rising gasoline and food prices," said Ernest Istook, a former congressman. "Who deserves the blame? Middle East sultans? Oil company executives? Commodities speculators?

Today the American people are paying high food and gas prices as secular progressive politicians try to shift the blame.

"How about blaming our very own United States Congress?" Istook asked. "For decades, congress has led our government into disastrous decisions by being the patsy of radical environmentalists, naysayers and prophets of doom. Recent presidents have done little to resist."[4]

According to Istook, we must reverse misguided federal polices in order to lower food and gas prices.

"The stifling of domestic oil and gas production and the suppression of new refineries and nuclear power plants have choked off the supplies of domestic energy, forcing us to rely on foreign oil," Istook said. "In the international market, we must bid against the growing energy appetites of China and India, and we're held hostage by the oil cartels of OPEC. The world market is unstable and expensive, and we shouldn't be at its mercy."

Istook pointed out that there really is no shortage of oil and gas in the United States but these great reserves have been placed off limits by Congress.

"The American Petroleum Institute... reports that opening up these areas would provide enough oil to power 60 million cars for 60 years... " Istook said.

Istook estimates that sixty percent of our oil is now imported and this has cost us one million jobs in the oil and gas industry in the past twenty-five years.

"Similar federal policies have blocked construction of oil refineries and nuclear power plants for more than 30 years, again increasing our dependence on foreign supplies of energy," Istook said.

Representative John Shadegg of Arizona also believes congress is to blame for the high cost of gasoline and other fuels. He says that for politically correct reasons, Congress "has locked up millions of acres of land in the western U.S., Alaska and the Outer Continental Shelf, where there is plenty of oil to be explored."

"The mandate set off a domino effect as the government pays farmers to grow corn rather than other grains, and to sell it for fuel (ethanol) instead of food," Istook said. "And because corn is the major feed for livestock, the prices of meat, eggs, milk and so on climb along with prices for grain, flour, baked goods, etc."

Istook believes it's time for action. He recommends that we open up our oil and gas reserve areas; build refineries and nuclear power plants; stop the expensive and wasteful mandates such as ethanol; and allow the American free enterprise system to develop alternative fuels for the future.

Remember Murphy's Law: everything that can go wrong will go wrong? I think that applies to our Congress.

Sometimes I think that the members of Congress are living on some other planet.

I was surprised to learn that the House of Representatives voted to honor Ramadan, the Islamic holiday.

The resolution was sponsored by Rep. Eddie Johnson of Texas and said the House recognized the beginning of Ramadan, the Islamic holy month of fasting and spiritual renewal, and commended Muslims in the United States and throughout the world for their faith.

Although it passed overwhelmingly, several representatives were less than enthusiastic about the resolution.

Why? Because they remembered that an attempt to get Congress to honor Christmas was met with strong opposition.

Former Congressman Tom Tancredo of Colorado said the Ramadan resolution was an example of how political correctness has captured the congressional and media elite in this land.

"I am not opposed to commending any religion for their faith," Tancredo said. "The problem is that any attempt to do so for Jews or Christians is immediately condemned as 'breaching' the non-existent line between Church and State by the same elite."

A few weeks after passing the resolution, nine members of the House refused to vote for a Christmas resolution that noted the holiday "is celebrated annually by Christians throughout the United States and around the world."[5]

Although the resolution was ultimately approved, the American Family Association noted that it was quite re-markable for the nine house members to vote against Christmas.

The "Upper House" or the United States Senate is also full of surprises.

Senate Majority Leader Harry Reid of Nevada, a rain-maker for the far left, blamed the most recent California fire on global warming. He made the statement on the floor of the United States Senate.

Now, let me ask you: How outrageous is this?

Secular progressives like Senator Reid never miss an op-portunity to further their agenda.

Meanwhile, Sen. Barbara Boxer of California claimed the war in Iraq was to blame for the slow progress in fighting the fire. She insisted that Gov. Arnold Swartznegger should have called out the National Guard to help the fire-fighters. But, alas, said Senator Boxer, the National Guard was in Iraq.

It always amazes me that the news media never chal-lenge these left-leaning liberals when they make those kinds of erroneous statements.

Here's the truth. There were two thousand California guardsmen in Iraq, but seventeen thousand still in California. A guard spokesman said that seventeen-hundred guardsmen were deployed to fight the fire and that fourteen thousand were still available.

This flatulent rhetoric from Senators Reid and Boxer would make a great episode for *Dumb and Dumber: With Harry and Barbara.*

Now here comes California then-Lieutenant Governor John Garamendi who echoes Boxer's irresponsible statement.

"How about sending our National Guard back from Iraq, so that we have those people available here to help us?" he asked.

He obviously made those statements to further his own political career and rally his radical anti-war base. While the people of Southern California were running from the wind-whipped firestorms trying to survive, find food and water and a place to sleep, and while forty firefighters were being injured, the three politicians used global warming, bashing then-President George W. Bush and the War in Iraq to push their personal left-wing agendas. However, firefighters on the scene said that had the state provided adequate aircraft and personnel earlier, the devastation could have been prevented.

"It is an absolute truth – had we had more air resources we would have been able to control this fire," said Orange County Fire Chief Chip Prather.

But Boxer and Garamendi blamed President Bush for their own failures. And that's wrong.

Yet the national news media never uncovered the real truth about the devastating fire.

I'm not clairvoyant, but I wonder if that is why fewer and fewer people are watching these irresponsible network news programs? The corruption in the media is disgraceful. It's off the chart.

We later learned that two-dozen helicopters and two massive cargo planes, equipped with large tanks of water, initially were not used because of state regulations. Those regulations require "fire spotters" from the state forestry department to accompany the pilots to show them where to drop the water.

Voila! Global warming was not the cause of the fire. Neither was the War in Iraq nor our embattled former president.

Give me a break! Any pilot could find the fire!

Hello! Neither the governor NOR the lieutenant governor took decisive action. And where was Boxer?

However, Swartznegger said that the heavy winds kept him from being able to send more planes in to fight the fire. He was correct. There has always been tension between nature and man.

The fire destroyed forty-six-thousand acres and fifteen-hundred homes and displaced a half million people. One of the nine-thousand firefighters said, "This ain't hell, but you can see it from here."

There was another problem that contributed to the fiery devastation. For years environmentalists, with their apocalyptic doom and gloom, have been fighting against property owners' rights to clear brush away from the lands and the state government's controlled burning. Most foresters know that underbrush fuels fires and that thinning the brush through controlled burning is a good method of fire prevention. But California environmentalists won victory after victory in the state legislature and were able to get the procedures banned.

Ellis Washington in The Report from Washington asks a very probing question: "Could there possibly be some public policy reasons why there are so many fires in California notwithstanding Senator Reid's preposterous global warming fairytales?"

"Radical liberal environmentalist policies prohibit controlled burning of heavily forested areas," he said. "Years and years of dormant undergrowth in the forests provide perfect kindling for a fire-prone environment like California."

Washington also pointed out that in that state the majority of the federal forest-thinning proposals have been tied up in costly litigation by environmentalists.

"The GAO (General Accounting Office) examined 762 U. S. Forest Service proposals to thin forests and prevent fires during the past two years," Ellis said. "According to the study, slightly more than half the proposals were not subject to third-party appeal."

Washington, taking a much-more common-sense approach, says the governor should insist that the California Legislature mandate "controlled forest burning, undergrowth removal and clear-cutting policies."

But obviously that would be far too simple for the ultra-liberal politicians in California and would put a lot of environmentalist hustlers out of work.

"In Roman antiquity, there was a prescient aphorism that stood for 2,000 years as a monument to bureaucratic arrogance, stupidity and ineptitude – 'Nero fiddled while Rome burned,'" Washington said. "In modern parlance, it could be iterated: As Senator Harry Reid spews out nitwit global warming propaganda, California burned."[6]

The California fire and Hurricane Katrina have taught us one salient truth: those who rely on government will be disappointed.

Now can you top this? Remember when the bridge over Interstate 35 collapsed in Minneapolis, Minneapolis, in early January of 2008? It was a terrible tragedy. Thirteen people were killed and one hundred forty-five injured when the bridge fell one-hundred-fifteen feet down into the Mississippi River.

All of us were stunned as we watched television news

reports of fire department divers trying to recover the bodies from the river.

Once again, Senator Reid used the tragedy to hammer President Bush.

Senator Reid said the president had been so distracted by the War in Iraq that he had neglected the infrastructures including roads and bridges in the land.

However, what was the real problem?

According to Chairman Mark Rosenker of the National Transportation Safety Board, because of design error, the connectors, called gusset plates, were only half the one-inch thickness they should have been, Frederic J. Frommer of the Associated Press reported.[7]

"Investigators found 16 fractured gusset plates from the bridge's center span," Frommer reported.

But Senator Reid continued to bash President Bush and the news media reported Reid's statements as though they were true.

However, I did my own investigation and learned that the bridge was constructed in 1967 when Lyndon Johnson was president.

I wonder if Senator Reid will ever retract his statements about the former president.

Sen. John Kerry of Massachusetts blames climate change for tornados that killed at least fifty people throughout the Southeastern United States in early 2008.

Kerry used the storms as a platform to advance his global warming agenda as he said the intensity of the storms is related to the warming of the earth.[8]

Has the former presidential candidate now become a weather man?

Kerry's statement that tornado activity is related to global warming has been questioned by one meteorologist with the Storm Prediction Center. "As of this writing, no scientific studies solidly relate climatic global temperature to trends in tornadoes," said meteorologist Roger Edwards.

One of the more ignominious actions of the United States Senate was when thirty-three senators voted against English as the official language in America.

After the vote, Colonel Harry Riley, U. S. Army, retired, wrote a letter to the senators and said, "I can only surmise your vote reflects a loyalty to illegal aliens."

"I don't much care where you come from, what your religion is, whether you're black, white or some other color, male or female, democrat, republican or independent, but I do care when you're a United States Senator, representing citizens of America and vote against English as the official language of the United States," he said.

Colonel Riley noted that four of those voting against English were then-presidential candidates: Senators Joe Biden, Hillary Clinton, Barack Obama and Chris Dodd.

"Those 4 senators (are) vying to lead America but won't or don't have the courage to cast a vote in favor of English as America's official language when 91% of American citizens want English officially designated as our language," the colonel said.

A Hindu chaplain from Reno, Nevada, by the name of Rajan Zed, delivered the opening prayer in the U.S. Senate in July of 2007. Zed told the *Las Vegas Sun* that during his prayer he would refer to ancient Hindu scriptures, including Rig Veda, Upanishards, and Bhagavard-Gita, according to Jim Brown of OneNewsNow.

David Barton, a Christian historian and president of WallBuilders, questioned the decision to invite Zed to lead the prayer.

"Barton points out that since Hindus worship multiple gods, the prayer will be completely outside the American paradigm, flying in the face of the American motto 'One Nation Under God,'" Brown said.

In his prayer, Zed said, "We meditate on the transcendental glory of the deity supreme who is inside the heart of the Earth, inside the life of the sky, and inside the soul of the heaven."

An American hero who received the Congressional Medal of Honor is speaking out against all the anti-military rhetoric coming out of Congress.

"American pilots do not conduct air-raids on villages, killing civilians, nor are our troops cold-blooded murderers or terrorists," said retired Air Force Colonel George Day. "But if you believe some current members of congress who have accused our men and women in uniform of all that and worse, you would have to conclude our military is a barbarian horde... The truth is that some in the U. S. Congress and their mouthpieces in the media now represent a much bigger threat to the lives of our men and women in combat, and our national security, than any foreign enemy."

Charley Reese, a veteran journalist writing in the *Orlando Sentinel-Star* , said, "Politicians are the only people in the world who create problems and then campaign against them."

"You and I don't propose a federal budget," he said. "The president does. You and I don't have the Constitutional authority to vote on appropriations. The House of Representatives does. You and I don't write the tax code. Congress does."

Reese said that the president, congress and the U. S. Supreme Court justices are responsible for most of the problems that plague this land.

"Those 545 human beings (president, Congress, high court) spend much of their energy convincing you that what they did is not their fault," he said.

He believes the people can take back the control of this

land.

"It seems inconceivable to me that a nation of 300 million cannot replace 544 people who stand convicted... of incompetence and irresponsibility," he said.

According to Reese, if the tax code is unfair it is because those in government want it to be unfair.

"If the budget is in the red, it's because they want it in the red," he said.

Reese cautions all of us not to believe there are mystical forces such as "the economy," "inflation" or "politics" that prevents those officials from doing what they have taken a solemn oath to do."

"Americans routinely elect and re-elect politicians who act in ways that should make us cringe," said Ben Shapiro in his column titled "The Death of Shame."[9] "We're engaged in a vicious cycle of degradation: Politicians act with utter disregard for common decency, Americans tolerate it; politicians act with greater disregard for common decency, Americans tolerate it even more. And the cycle goes on."

George Orwell said the time would come when "Big Brother" would be watching all of us and, of course, he was talking about big government. But I wonder who is watching Big Brother?

Chapter Six
Who's In Charge?

As America staggers down a crooked road, sometimes I wonder who's charge when:

- Homeland Security distributes taxpayer cash for bingo, limousine service and other non-security-related projects.
- Busy airport security screeners fail to detect "fake bombs" in security-test stings.
- A nuclear warhead is mistakenly loaded on board a B-52 Bomber without the knowledge of the pilot or crew.
- Nuclear plant guards are caught sleeping.
- Sabotage, drinking reported at NASA.
- There is a drunken brawl at the U. S.-operated research facility in Antarctica.
- Federal spending to rebuild the City of New Orleans after Hurricane Katrina is a disaster.
- There are reports of lax and lazy security at the Los Alamos nuclear-weapons laboratory.
- Imprisoned sex offenders receive thousands of dollars in federal grants.
- The U. S. Treasury Department plans to send $496.5 million to the Palestinian Authority now controlled by the Hamas terrorists committed to the destruction of Israel.

When I hear of failures like these, I wonder if the federal government is out of control or so large and encumbered with bureaucratic red tape that it is unmanageable.

We see government blunders and mismanagement everywhere.

A survey of Homeland Security grants to state and local governments is quite revealing and disturbing. Some of the money went to purchase a bus, provide for a bingo hall and a limousine service, according to a report from the Heritage Foundation.

The report also revealed that security funds were used to build a homeless shelter and to fund a program to locate missing persons.

What does the bingo hall, limousine service and the other projects have to do with homeland security? And again, I ask, "Who's in charge and how could this happen?"

I'll admit it's a good deal for some of our citizens. Now if terrorists detonate a bomb in their city, it will undoubtedly shut down the bingo game but the politicians will have a bus and a limousine to get them out of town.

Obviously, Homeland Security officials do not agree with the report.

Why does that not surprise me?

Now for some more exciting news.

Several years ago, two of our nation's busiest airports, Los Angeles International and O'Hare in Chicago, failed to locate "fake bombs" during a screening sting.

The bombs were carried onto several airplanes by undercover federal agents posing as passengers.[1]

Screeners at the Los Angeles airport missed about seventy-five percent of simulated explosives and bomb parts the agents hid under their clothes or in carry-on bags at various checkpoints. Security at O'Hare was also disappointing.[2]

The failure rate stunned the experts!

"Terrorists bringing a homemade bomb on an airplane, or bringing on bomb parts and assembling them in the cabin, is the top threat against aviation," Thomas Frank of

USA Today reported.

Homeland Security's response was typical. They are trying to find better methods to protect the public, whatever that means.

Thankfully, they now have instituted far better security checkpoints at all of our major airports.

A B-52 Bomber, mistakenly loaded with five Advanced Cruise Missiles with nuclear warheads, flew from Minot Air Force Base in North Dakota to Barksdale Air Force Base in Louisiana. The pilot and crew didn't have a clue they were carrying the dangerous warheads.[3]

Michael Hoffman of the *Military Times* reported that since the warheads were not discovered until the B-52 landed at Barksdale, they were unaccounted for during the three-and-one-half hour flight.

Of course, when I heard about the incident, I wondered what would have happened if the B-52 had crashed somewhere along the route from North Dakota to Louisiana.

"A crash could ignite the high explosives associated with the warhead, and possibly cause a leak of the plutonium, but the warheads' elaborate safeguards would prevent a nuclear detonation from occurring," an Air Force spokesman said.

I sure hope so.

There's more!

Have you ever heard of the Peach Bottom Nuclear Plant in Pennsylvania? It is one of America's largest.

CBS News reported that several security guards at the plant went to sleep on the job.

"The area where the guards were taped sleeping on different shifts and days, is called the 'ready room,'" CBS reported. "The sleeping guards are supposed to be poised to spring into action immediately if there is an emergency."

I take that to mean the guards are to be on "alert" in case there is a nuclear meltdown or a so-called "China Syndrome."

However, a spokesman for Exelon Nuclear, operator of Peach Bottom, said the sleeping guards did not "impact the safety and security of the plant."

Really? I'm sure that was a great relief to the Pennsylvania people living near the plant.

I guess good help is hard to find these days.

Did you hear the story about the NASA astronaut charged with attempted murder in a bizarre love triangle?

Astronaut Lisa Marie Nowak was accused of attempted first-degree murder in a bizarre attack on a romantic rival for a space shuttle pilot's affections.[4] Nowak, a navy captain, was a forty-three-year-old mother of three.

According to a police report, Nowak drove nine hundred miles to confront Colleen Shipman, a woman she believed was trying to win the affections of Navy Commander William Oefelein, an unmarried fellow astronaut.

When she learned that Shipman has planned a trip to Orlando, Florida, Nowak followed her.

"Nowak raced from Houston to Orlando wearing diapers in the car so she wouldn't have to stop to go to the bathroom..." Fox News reported. During space flights, astronauts wear diapers during launch and re-entry.

According to police, when Nowak arrived in Orlando, she went to the airport to wait for Shimpan's plane to arrive. She was wearing a wig and a trench coat. Then Nowak followed Shipman to her car where she attacked her.

Shipman survived and was able to call the police. When they arrived they found a steel mallet, a four-inch folding knife, a can of chemical spray and six-hundred dollars in a bag Nowak was carrying.

Nowak is a graduate of the U. S. Naval Academy at Annapolis, Maryland, and has a master's degree in aeronautical engineering. She flew to the International Space Station in July of 2006 aboard the space shuttle Discovery.

She was charged with attempted kidnapping, burglary

and assault and battery. She entered a guilty plea to a lesser charge and she received a year's probation.

There were two other reports out of Cape Canaveral that shook the foundations of the Space Agency.

"One involved claims that astronauts were drunk before flying," Marcia Dunn of the Associated Press reported. "The other was news from NASA itself that a worker had sabotaged a computer set for delivery to the international space station."[5]

Apparently on two occasions astronauts were allowed to fly even though flight surgeons and other astronauts said they were drunk and a safety hazard.[6]

At a news conference, Bill Gerstenmaier, NASA's space operations chief, was asked repeatedly about the drunken astronaut report. He replied he had never seen an intoxicated astronaut before a flight.

But Gerstenmaier confirmed the computer sabotage.

"He revealed that an employee for a NASA subcontractor had cut the wires in a computer that was about to be loaded into the shuttle Endeavor for launch," Dunn reported.

The operations chief declined to identify the worker, the subcontractor or where the incident took place. He said the computer did not pose a safety problem and that it would be repaired.

Those of us who were eyewitnesses to the rise of NASA and the work of our heroic astronauts always looked on them with great pride.

Have we now witnessed the rise and fall of this great institution? I hope not.

Christmas always brings out the best in some people, the worst in others.

Two members of an Antarctic base staff were evacuated to Christ Church, New Zealand, from the most remote research facility after a drunken brawl, according to the independent British newspaper *Guardian Unlimited*.

The fight took place at the U. S.-operated station located in the heart of the frozen continent where scientists carry out a range of investigations from astrophysics to seismology.

And the nonsense goes on and on.

There was a serious security breach at Los Alamos, the nation's premier nuclear-weapons laboratory in New Mexico.

First, we learned of a leak of highly classified information on the Internet. Later, a worker took his laboratory laptop computer, with additional sensitive information, on a vacation to Ireland where it was stolen from his hotel room, according to John Barry of *Newsweek*.[7]

"It has not been recovered," Barry reported.

Another breach at the same lab involved a scientist who worked in the experimental physics division involved with weapons design. It occurred when he sent an email, containing secret information, to employees at a Nevada test site over the Internet rather than through an internal secure network.[8] Of course, anyone could access the information.

"These incidents come as Los Alamos is still reeling from the revelation that, in January (2007) half a dozen board members of the company that manages the lab circulated – over the Internet – an email to each other containing... information about the composition of America's nuclear arsenal," Barry reports.

So what do we hear from Los Alamos officials? "The purported incident is under investigation; it would be inappropriate to comment."

Typical bureaucratic doublespeak!

The Associated Press (AP) has reported that dangerous sex offenders receive thousands of taxpayers' dollars to take college courses by mail.[9]

"Across the nation, dozens of sexual predators have been taking higher education classes at taxpayer expense while confined by the courts to treatment centers," wrote Ryan J. Foley for AP. "Critics say they are exploiting a loophole to receive Pell Grants, the nation's premier financial aid program for low-income students."

Foley said that even though prison inmates and students convicted of drug offenses are not eligible for the Pell grants, sexual predators can lawfully receive the grants if they are transferred from prisons to state or federal institutions for treatment.

"This is the most insane waste of taxpayer money that I have seen in my eight years in Congress," Said Rep. Ric Keller of Florida who in 2008 said he wants to stop the practice. "It is a national embarrassment that we are wasting taxpayer dollars for pedophiles and rapists to take college courses while hardworking young people from lower-class families are flipping hamburgers to pay for college."

Officials at some of the treatment centers have reported the sex offenders have used the Pell grant funds to purchase DVD players CDs and clothing after they have dropped out of school.

"The institutions and the government do not keep count of how much money sexual predators receive," Foley said. "The maximum Pell Grant is $4,310 per year. The government generally sends payments to colleges for tuition, and any leftover is sent to the student to cover expenses."

Here are some examples of the predators receiving taxpayer funds:[10]

- In early 2008 six predators at the Sand Ridge Treatment Center in Wisconsin were receiving Pell Grants and others at that institution received the grants in the past. Sand Ridge administrators said some of the predators used the money for living expenses already being provided by the taxpayers of Wisconsin.

- In Iowa, fourteen convicted sexual predators at the Cherokee Mental Health Institute have received grants during the past few years. However, nine of them dropped out of the program after receiving the government taxpayer funds.

- In California, several of the predators living at the Coalinga State Hospital are receiving government funds but a representative of the institution said they have no way of determining how many or how much they receive.

Some members of Congress say it would be wrong to stop the predators from receiving the grants since they help in the rehabilitation of the predators.

Back closer to home, congress exercises certain jurisdictional authority over Washington, D. C. but has never done a very good job.

Some areas of the city look like a war zone.

The most recent congressional failure is the use of tax dollars to provide clean needles to drug addicts in the city. Officials will spend one-million dollars on the program in an effort to reduce the AIDS epidemic.

Now, on the surface, that sounds like a good program. But let's take a closer look. The funds subsidize bad behavior, both immoral and drug related.

Gary Bauer, president of American Values, questions the program since there is no available research to indicate it will be successful.

"The problem is not dirty needles, the problem is not unprotected sex," Bauer says. "The problem is that large numbers of people in urban areas continue to engage in behavior that is guaranteed to spread disease…"

He also says the program subsidizes "bad behavior" – intravenous drug usage and the homosexual lifestyle – and that it is "outrageous" the government is subsidizing that kind of immorality.

There are times when I wonder if there is any common

sense left in Washington.

For instance, our government has given hundreds of millions of dollars to the Palestinians, a blood-thirsty gang committed to wiping Israel off the face of the earth, according to columnist Linda Chavez writing for the *Jewish World Review.*

Chavez rightly points out that we should have learned a lesson from the American experience that handouts do not work. Rather they become a crutch and create a permanent underclass dependent on government.

"At least American welfare recipients weren't using the cash… to build bombs to blow up their neighbors," Chavez wrote in a commentary entitled "Palestinian Handouts." "The same can't be said of the Palestinians who have received Western aid. Israelis have found a bomb-making factory in the West Bank town of Nablus."

Chavez also says the Palestinians have built a culture based on hate and this country shouldn't give them another dime until Hamas and Fatah quit fighting among themselves and trying to destroy Israel.

In 2008 the United States planned to give $496.5 million to the Palestinian Authority controlled by Hamas terrorists. Of that amount, $410 million is for development of programs and an additional $86.5 million for "security training" of Palestinians by the CIA.[11]

Since 1994, the CIA has armed and trained thousands of the security forces and many of them later joined various Palestinian terrorist organizations.[12]

"CIA Palestinian training success is best described by a member of the PA's own security unit Force 17 officer Abu Yusef: 'The operations of the Palestinian resistance would [not] have been so successful and would not have killed more than 1,000 Israelis… and defeated the Israelis in Gaza without [American military] training,'" he boasted.[13]

Apparently there is no way for the United States to insure the Palestinians use America's millions of dollars for humanitarian purposes.

In March 2007, PA Prime Minister and former World Bank official Salam Fayyad, told London's *Daily Telegraph*: "No one can give donors that assurance that funds reach their designated destinations."

Hence, no one including the State Department of the United States knows for sure how much of this country's aid to the Palestinians winds up in the hands of terrorists.

Perhaps some of the American dollars goes to Al-Aqsa television, an anti-Israeli, anti-American station that telecasts Tomorrow's Pioneers, a children's program.

The station once featured Assoud the Rabbit who told the young kids watching the program he would "annihilate the Jews."

They also presented a puppet show that features a valiant young Palestinian boy who breaks into the White House in Washington, D.C. and kills the American president, but not before making him beg for his life.

I get quite indignant when I realize that taxpayer dollars are funding that kind of foolishness. Sure, we have the secular progressives in the Jimmy Carteresque tradition telling us that all we have to do is sit down with Hamas and other terrorist-oriented groups, shake their hands and everything will be alright.

Someone should stand up and say, "Enough is enough!" But no one seems to think it's that important. Perhaps that's because they aren't spending their own money to help the Palestinians kill Jews. They're spending our money.

Our government is woefully inept and nowhere is that more evident than in the efforts to restore New Orleans and the Gulf Coast after Hurricane Katrina.

Representative Tancredo once said it is "time the taxpayer gravy train left the New Orleans station" and called for an end to federal aid to the area that was devastated in

2005.

There are reports that the federal government has spent one-hundred-sixty-one-billion dollars to rebuild much of New Orleans and the Gulf Coast.

Tancredo pointed out that an estimated one-billion dollars has been wasted through fraud and abuse in the recovery program.

All of us were astounded when we learned that the trailer houses provided by the Federal Emergency Management Administration for homeless Katrina refugees were toxic and making the people sick and some died.

"Along the Gulf Coast, in towns and fishing villages from New Orleans to Mobile, survivors of Hurricane Katrina are suffering from a constellation of similar health problems," Amanda Spake wrote in *The Nation*. "They wake up wheezing, coughing and gasping for breath, their eyes blurry; their heads ache; they feel tired, lethargic. Nosebleeds are common, as are sinus infections and asthma attacks. Children and seniors are most severely affected, but no one is immune."

What caused the problem? The 102,000 travel trailers and mobile homes purchased by FEMA are built with composite wood and particle board and other components that emit formaldehyde, a toxic chemical, according to Spake.

After the devastation, some 275,000 homeless people were living in the trailers that cost taxpayers nearly three billion dollars.

Formaldehyde is a powerful irritant that when inhaled constricts breathing passages. It has also been classified as a carcinogen.

FEMA knew for more than a year that the trailers were toxic before they started moving people out of the trailers.

Again I ask: Who's in charge?

Reverend Jesse Lee Peterson, a black writer with a passion for the truth, says there have been myriad misconceptions about the recovery program pawned off on the

American people by agenda-driven politicians, the frantic news media, desperate to criticize President George W. Bush, and by "race hustlers."[14]

Peterson points out that although government officials in Washington were responsible for the toxic trailers fiasco, the people of New Orleans and their leaders must take the blame for many of the problems after Katrina.

"If you're black and a hurricane is about to destroy your city, then you'll probably wait for the government to save you," Peterson said. "When 75 percent of New Orleans residents had left the city, it was primarily... welfare-pampered blacks that stayed behind and waited for the government to bail them out."

According to Peterson, men like Reverend Jesse Jackson and Nation of Islam leader Louis Farrakhan condemned the "racist" President George W. Bush for the fiasco in New Orleans after Katrina.

"Farrakhan actually proposed the idea that the government blew up the levee so as to kill blacks..." Peterson said.

But he believes that if Farrakhan and Jackson were really serious about blaming government, they should blame the City of New Orleans and the State of Louisiana.

"Responsibility to perform – legally and practically – fell first on the mayor of New Orleans," Peterson said. "We are now familiar with Mayor Ray Nagin...who likes to yell at President Bush for failing to do Nagin's job."

The Washington Times reported that Nagin even failed to follow the city's emergency-response plan which pointed out that thousands of the city's poorest people would have no way to evacuate the city during a hurricane.

According to Peterson, there were ways to evacuate the poor. "We have photographic evidence... showing 200 parked school buses, unused and under water," Peterson said. "How much planning does it require to put people on a bus and leave town...?"

Instead of launching an evacuation, Mayor Nagin crowded tens of thousands of people into the Superdome and the city's Convention Center. After only three days, those facilities became ghettos, rampant with theft, rape and murder.

These are only a few of the hundreds of examples of the paralysis that plagues federal government and although there are some voices calling for reform, it may be too little too late.

We the People elect a president, vice president, senators and congressmen who are supposed to be looking out for our interests. Yet there is a cavalier attitude about handling our tax dollars that defies description as liberal democrats and marshmallow republicans think of hundreds of ways to spend the money we loan them.

Now, welcome to the world of waste!

Missing funds, government employee credit card embezzlement, student loans for fictitious students attending fictitious colleges, fraudulent government research, uncollected airplane ticket refunds and massive program duplication total $100 billion dollars a year!

That's the word from a study by the Heritage Foundation, a prominent conservative think-tank.

It makes me want to shout: "Does anyone care about us little guys who are paying the bills?"

In 2003, the Treasury Department reported that $24.5 billion was missing! These are funds for which government auditors cannot account.

"The government knows that $25 billion was spent by someone, somewhere, on something, but auditors do not know who spent it, where it was spent, or on what is was spent," the Heritage Foundation reported.

According to Heritage, an audit revealed that during one five-year period, the Department of Defense alone pur-

chased and then left unused about 270,000 commercial airline tickets at a cost of one-hundred-million dollars. Auditors also found twenty-seven-thousand transactions where
the Pentagon paid two times for the same ticket.

"Even worse, the Pentagon never bothered to get a re-
fund for these *fully refundable* tickets," the report noted.

There was another very remarkable practice concerning
the airline tickets. The Pentagon purchased tickets for
flights for their employees then for some unknown reason
reimbursed the employees for the costs of the same tickets.

"In one case, an employee who allegedly made seven
false claims for airline tickets professed not to have noticed
that $9,700 was deposited into his/her account," the report
said. "These additional transactions cost taxpayers $8 mil-
lion."

The Pentagon announced it would build a new eight-
een-hole golf course near Andrews Air Force Base in suburban Maryland at a cost of about of $5 million dollars.
However, *Golf Digest* reported there were already nineteen
military golf courses around Washington, D.C.

Then there is the case of embezzled funds at the U. S.
Department of Agriculture (USDA). The Department issued federal credit cards to their employees to purchase
job-related materials to be paid for by the agency. But large
numbers of those employees abused the credit card privilege.

"A recent audit revealed that employees... diverted millions of dollars to personal purchases through their gov-
ernment-issued credit cards," the report said. "Sampling
300 employees' purchases over six months, investigators
estimated that 15 percent abused their government credit
cards at a cost of $5.9 million."

Taxpayer funds were used for tickets to an Ozzie Os-
bourne rock concert, tattoos, lingerie, car payments, cash
advances and even tuition to bartender school.

Department officials have pledged a thorough investigation but following the missing dollars may prove to be quite elusive. There are 55,000 credit cards in circulation, including 1,549 in the hands of former employees who no longer work at the USDA.

The Agriculture Department also spent a billion dollars in farm payments to more than 170,000 dead people over a seven-year period, according to congressional investigators.

The Defense Department also uncovered a credit card scandal.

"During one recent 18-month period, Air Force and Navy personnel used government-funded credit cards to charge at least $102,400 for admission to entertainment events, $48,250 for gambling, $69,300 for cruises, and $73,950 for exotic dance clubs and prostitutes," the Heritage report said.

However, Medicare overspending is one of the most corrupt of all federal programs.

"Medicare wastes more money than any other federal program, yet its strong public support leaves lawmakers hesitant to address program efficiencies, which cost taxpayers and Medicare recipients billions of dollars annually," the report said.

Auditors discovered that Medicare pays as much as eight times what other federal agencies pay for the same drugs and medical supplies. They learned that in some cases, Medicare paid an average of more than double what the Veterans Administration paid for the same items.

However, inflated prices for drugs are not the most expensive examples of Medicare waste.

"Basic payment errors – the results of deliberate fraud and administrative errors – cost $12.3 billion annually," the auditors said. "As much as $7 billion owed to the program has gone uncollected or has been written off... Putting it all together, Medicare reform could save taxpayers and program beneficiaries $20 billion to $30 billion annually without reducing benefits."

Some estimate that amount in 2010 has escalated to fifty-billion-dollars annually.

There is also significant waste and graft in the Department of Education particularly as it relates to funding fictitious students at fictitious colleges.

For instance, the department received applications for student loans from three students to attend the Y'Hica Institute in London, England. The Department approved the applications and disbursed $55,000 dollars.

"The Education Department administrators overlooked one problem: Neither the Y'Hica Institute nor the three students who received the $55,000 existed," the Heritage report said. "The fictitious college and students were created (on paper) by congressional investigators to test the… verification procedures."

Recent studies reveal that nearly $22 billion dollars in student loans have never been paid back to the Department.

Does anything shock the American people these days? Have we become so accustomed to being ripped off that graft and fraud no longer even bother us? Or have we been inoculated against the anger that should result from bad government behavior?

Now let's look at some other outrageous cases of government waste.

The government spent nearly $20 million on the International Fund for Ireland, whatever that is, for projects including pony-trekking (riding across country on ponies) and golf videos.

Congressmen always look out for themselves. They have fully automated push-button elevators on Capitol Hill but they continue to spend thousands annually to have attendants push the buttons for them.

"The Pentagon and Central Intelligence Agency

channeled some $11 million to psychics who might provide special insight about various foreign threats," Thomas A Schatz of Citizens Against Government Waste reported. "That was the disappointing 'Stargate' program."[15]

Since the bureaucrats in Washington are playing with our money, there are more wild spending schemes that anyone can even imagine.

The Social Security Administration spends $25 billion a year on Supplemental Security Income. Studies reveal:[17]

- Thousands of men and women in prison receive SSI checks related to their alleged disabilities.
- In Denver, $160,000 was sent to several recipients as their "official address" which was a local tavern.
- A San Francisco drug addict uses his SSI income to buy drugs to sell to other addicts on the streets for a profit.
- Estimates suggest that as many as seventy-nine-thousand alcoholics and drug addicts in the land receive SSI benefits, about $360 million each year, and spend the money on their habits.

Chapter Seven
The Snake Pit

Hollywood is another indelible indicator of a culture in decline as today's movies baptize the American people in all kinds of profanity and gratuitous sex.

Through the years, the once-proud movie kingdoms such as MGM, Warner Brothers, Universal International, Republic and others produced dramas of epic greatness that kept us all spellbound, dramas such as *Casablanca, Citizen Kane*, the *Ten Commandments, Ben Hur, Gone With the Wind* and hundreds of other memorable movies.

However, today Hollywood is a snake pit that defines morality lost with its obsession with profanity, immorality and violence that is like a poison to our nation.

Washington, D.C. is America's great center of power, but Hollywood is the center of influence, a vast opinion-molding empire that shapes America's values more than any other institution in our land.[1]

"And yet, Hollywood's 'beautiful people' – the celebrities as well as the movers and shakers behind the scenes – share a worldview both alien and poisonous to traditional America," says David Kupelian, managing editor of WorldNetDaily. "With rare exceptions, Hollywood celebrities always seem to champion outrageous or immoral positions in crucial national issues, and to aggressively use their social power and prestige to advance such agendas."

Irreverent and blasphemous remarks are common in current movies and television programs and the American people, including hosts of Christians, seem mesmerized by it all.

One year Kathy Griffin won an Emmy for the best reality series "My Life on the D-List."

During her irreverent acceptance speech at the Television Arts and Sciences ceremony, she used profanity and even blasphemed the name of Jesus.

She noted that when most stars receive their awards they thank God.

"I want you to know that no one had less to do with this award than Jesus," she said. "This award is my god now."[2]

Earlier, Griffin described herself as a "complete militant atheist."

Griffin's obscenities and vulgar slur against Jesus should raise righteous anger in the hearts of Christians everywhere.[3] But most Christian leaders were silent on the Kathy Griffin affair as they are on most all moral and ethical issues.

A few days later, Larry King interviewed Griffin on CNN. At the close of the program, King told her, "You're the greatest!" What does that say about the unknown Larry King?

Although America is basically a Christian nation, Hollywood is anti-Christian. It appears to be obsessed with the dark side of life and is quite successful at making evil look innocent and bad, good.

Consider the ever-popular Harry Potter movies based on J. K. Rowlings' best-seller Potter series.

A close look at the content of her books and the movies that followed reveals a heavy emphasis on witchcraft. Here are some of the Potter movie themes: reincarnation and contact with the dead and spirit world; sorcery; divination; spells; curses; occult; black magic; demon possession; the dark aspects of witchcraft; and much more.[4]

The premiere episode of *Saving Grace* starring Holly Hunter was blatantly anti-Christian in content with an irreverent religious element.[5]

Here's the storyline: Hunter played an atheist detective by the name of grace whose life was characterized by alcohol, drug abuse and loose morals. While in a drunken stupor, she hit and killed a man with her car. A tobacco-chewing angel appeared and offered to help her recover from her moral collapse.[6]

"The primary content of the program focused on mocking God and attacking Christianity," said Ed Thomas writing for *OneNewsNow.*

Thomas adds the program was laced with explicit scenes of immorality, obscene language and alcohol abuse.

The American Family Association (AFA) pointed out that there were 55 instances of profanity in the opening episode, including numerous variations of taking God's name in vain.

AFA's Cindy Roberts notes that in the program Grace demonstrated no repentance, even though an angel supposedly was there to help her and that "gives the impression there is no repercussion when individuals intentionally sin."

Amazing Grace, now cancelled, was only one of dozens of profane, immoral, anti-Christian programs coming out of Hollywood, programs that ridicule churches, make fun of preachers and seek to convince the viewing public that Christianity is no longer relevant in this secular progressive culture. Thankfully, the program was cancelled.

By the way, this trashy program was on occasion sponsored by AT&T.

An episode of the CBS crime show *Cold Case Files* blatantly characterized devout Christian teenagers who favored abstinence before marriage as sexually active hypocrites who punished a member of their group in order to keep their sins secret.[7]

Pat Boone recalls that when he was a star in Hollywood, the studio executives aimed their pictures at the vast "fam-

ily audience," the fathers and mothers who took their children to the movies on Friday and Saturday nights.

"Of course, we hoped to appeal to the teens, but we
were aware that real success depended on our entertaining
the whole age range – and certainly not offending parents
by 'going over the line,' sexually or morally in any way.
No nudity, no profanity, no brutality and absolutely no explicit, graphic sex scenes," Boone said.

Boone points out that today Hollywood seems to be obsessed with the "soapy, sexy women's stories" they bring
to the big screen.

"Or they give us all-out soft porn – morbid, taboo-bust-
ing and hopelessly dark productions clearly intended to
appeal to every base instinct known to man," Boone said.

He calls today's movies "weapons of mass destruction."

According to Boone, there are scores of films that "undermine the honor and integrity of... our military, our
business leaders, and even Christian ministers..."

Columnist Bret Prelutsky, a former screenwriter for
M.A.S.H., agrees with Boone.

"For several years, I have marveled at the arrogance
Hollywood has displayed toward its customers..." Prelutsky said. "Why, I've been asking myself, have they insisted
on churning out one movie after another about lowlife
drug addicts? The only druggie movie I can think of that
made money was 'The Man With the Golden Arm,' back in
1955, and that was based on a best-selling novel and
starred Frank Sinatra and Kim Novak."[8]

Now, Prelutksy says, Hollywood has added a second
genre "that has proven equally unprofitable."

"I refer to war movies that present the American military in a bad light," he said. "No sooner, it seems, does 'In
the Valley of Elah' bomb at the box office than Hollywood
rolls out 'Redacted.' Even the titles they stick on these tur-
keys seem to have no other purpose than to keep audi-
ences as far away as possible."

The American people have an out-of-control obsession with celebrities. Otherwise, why would hundreds of people gather outside the New York City apartment of Heath Ledger after he died of a drug overdose?

I sense that millions of people live in pursuit of pleasure and they find it in the fantasy world of movies, perhaps in Heath Ledger movies. In such a fantasy world, the ordinary movie fan can vicariously travel to the Riviera on a private jet with their favorite stars, attend gala events with the so-called "beautiful people," and ride the Rockies and the Old West with two cowboy lovers. Or they can party at chick Hollywood night spots where they dance to raunchy rap music laced with four-letter words and decadence calling for violence against women and death to cops.

Teenagers in particular idolize the young lions and ingénues of Hollywood and want to be like them. They pay little attention to the dark side of that lifestyle with the excessive drinking and DUIs, the car wrecks and jail sentences, the drug overdoses and death.

Are the young stars a privileged class? I hardly think so. But they live with the arrogance that fame, fortune and adulation bring into their lives. They make their own rules and decide for themselves what is right and wrong.

Young America's obsession with movie stars has no limits. But who really cares if Britney Spears walks into a gas station and pretends to be British or if Paris Hilton tells the world she is a born-again Mother Teresa?

It's obscene to reward bad behavior, particularly when someone is as inconsequential as Paris Hilton or Britney Spears. Yet their escapades have hypnotized the whole country. What does that say about us?

The Parents Television Council (PTC) issued a report that documented the destructive influence MTV and Black Entertainment Television have on the young people of our generation. They discovered that daytime music video programming on these two channels features sexual, violent, profane and obscene incidents every 38 seconds.

"Being in the trenches fighting for better indecency enforcement and cable choice on behalf of millions of American families, we thought we'd seen it all," said Tim Winter, PTC president. "but even we were taken aback by what we found in the music video programs on MTV and BET..."[9]

Their target? Impressionable children.

"PTC found more than 1,300 instances of offensive content in a mere 14 hours of programming in March (2008), primarily in the form of sexually-charged images, explicit language, violence, drug use and sales, and other illegal activity," said Jim Brown writing for OneNewsNow.

Delman Cates of the Enough Is Enough Campaign of Maryland, the organization that commissioned the study, pointed out there is a serious problem in American popular culture including the "celebration and glorification of images that sexually objectify women, that glorify violence and criminal activity, that negatively stereotype black and Latino men as pimps, gangsters, and things – and also that market adult-themed material to children and youth."

The Campaign members have demonstrated outside the homes of the CEOs of BET and MTV and Viacom which owns both of the cable channels. They also plan to target the companies that pay for the offensive programming through advertising. Some of the companies include Wal-Mart, McDonald's, Procter & Gamble, Sony, AT & T and others.[10]

Is Hollywood a dream come true or a nightmare?

Certain celebrities are the epitome of *chutzpa* and adult onset narcissism and are lowering their moral standards more every year. They have become the vanguard of the counter-cultural revolution we see all around us.

Other celebrities believe they are authorities on just about everything in American life and politicians in Washington often invite them to testify as "expert" witnesses before congressional committees. Here are some examples:

- Actress Meryl Streep appeared on Capitol Hill and spoke against the use of Alar, a chemical sprayed on apples.[11] She told the congressional committee that when Alar gets hot it turns into something like rocket fuel and is harmful to children.
- Angelina Jolie visited Congress as a goodwill ambassador for the United Nations Commission on World Refugees.[12]
- Model Christy Brinkley urged congressmen to prevent nuclear accidents.[13]
- A congressional task force on agriculture invited actresses Jane Fonda, Jessica Lange and Sissy Spacek to speak on problems facing American farmers. "Two of them choked back tears as they denounced President Ronald Reagan's farm policies," said Michael Bates, writing for *The Reporter* Newspapers.

What were their credentials for giving advice to our congressman concerning agriculture? All three had played roles as farm women in movies and, hence, considered themselves experts on the subject.

Both the broadcast and print media for months warned the American people about the dangers of Alar and how it could cause cancer in children.

Three years later the American Medical Association said that Alar, when used properly, would cause no harm to anyone or to Meryl Streep who became known as "The Alar Queen."

Marion Cotillard of France won the Oscar for best actress in 2008 for her role in *La Vie En Rose* during the annual celebration of Hollywood self-love.

After the Oscars, we learned she has something in common with actress Rosie O'Donnell and actor Charlie Sheen. All three believe in a conspiracy theory that American government agents purposely bombed the Twin Towers in New York on 9/11.

During an interview in her homeland, she also questioned whether American astronauts actually walked on the moon.

"Did a man really walk on the Moon?" she asked in an interview with *Paris Premiere-Paris Derniere*. "I saw plenty of documentaries on it, and I really wondered. And in any case I don't believe all they tell me, that's for sure."

It makes one wonder what planet the 32-year-old actress is living on.

There are times when I wonder if Hollywood stars are under a curse of some unknown origin.

Judy Garland of "The Wizard of Oz" fame was America's sweetheart – talented, beautiful, perfect.[14]

Millions of fans idolized this beautiful and gifted young lady and were shocked to learn she had died of a drug overdose. She was only 47.

"Judy Garland had struggled for two decades with drug-and-alcohol addiction and had been married five times, was plagued with self-doubt and had made several suicide attempts," said Columnist David Kupelian.

There were other tragic stars such as Elvis Presley, the undisputed King of Rock and Roll, and Marilyn Monroe.

"In her day, Monroe was idolized as America's reigning 'sex goddess,'" Kupelian said. "Yet inwardly she grew increasingly conflicted and depressed, finally dying at 36 of a sleeping pill overdose. The recent death of Monroe wannabee Anna Nicole Smith was eerily reminiscent of her idol's tragic demise."

Millions of fans all over the world idolized Elvis Presley, a man with a velvet voice and unbelievable wealth. Yet he faced inner demons that he could not control such as drug addiction and times of depression that led to his isolation in his home at Graceland in Memphis, Tennessee. His excessive use of drugs finally killed him. He was 42.

A physician testified he had prescribed thousands of pills during the last year before Elvis died. They included amphetamines, barbiturates, narcotics, tranquilizers and sleeping pills.

Kupelian points out that for every superstar bent on self-destruction, there are hundreds of others filled with conflict and live dysfunctional lives.

"Drug and alcohol abuse are commonplace..." he said. "Yet these people possess everything most of us secretly covet – talent, fame, good looks, wealth, adoration."

What is the secret curse that affects so many stars? Kupelian believes there is a "powerful dynamic" common to most celebrities which, in essence, becomes a curse to them. It is their insatiable egos.

Andrew Breitbart and Mark Ebner explain in their ex-pose "Hollywood Interrupted," "This often lucrative self-obsession invariably becomes downright pathological..."

"Massive egos and narcissism may be the primary in-gredients for achieving and maintaining Hollywood suc-cess, but they are also the No. 1 cause of the grandiose foibles in their storied, disastrous personal lives," Breitbart and Ebner said.

They also point out that "the desire to become a star re-quires an incredible appetite for attention and approval."

"The problem is, living off approval and applause, and deriving your sense of self-worth from the praise of others, may feel great, but it also produces great problems," they said. "When it comes to being worshipped, human beings don't make very good gods, something the Good Book warns about repeatedly."

At times stars act like a bunch of spoiled children.

Pop star Justin Timberlake, on a world tour in Goteborg, Sweden, spit on teenagers, poured water on them from the balcony of his hotel room and verbally abused them.

But the impressionable Swedish teenagers, like those in America, overlook his and other stars' bad behavior and continue to worship the Justin Timberlakes of this world.

Although Elton John is not a Hollywood star, he usually hosts a party for many of them each year following the Academy Awards.

While on his way to perform for the Concert for Princess Diana in London after her untimely death, a policeman stopped his car to let another car, carrying Princes William and Harry, pass.

John flew into a rage, for which he is famous, and screamed at the policeman, "Get out of my way... Don't you know who I am?"

Yeah, the policeman knew but didn't care and made him wait until the car with the young princes passed.

That's my kind of cop!

Stacey Smith and Erick Johnson, writing for *Netscape*, have compiled a list of their candidates for the Most Outspoken Stars in Hollywood.

They include Martin Sheen, the tireless peace activist who criticizes most everything about America.

"The West Wing star has been arrested several times, beginning in 1986 at an anti-SDI (President Ronald Reagan's Star Wars initiative) demonstration he calls the happiest day of his life," Smith and Johnson wrote.

When he once addressed a conference in Canada, he said, "Every time I cross the border, I feel I've left the land of lunatics... I always feel a bit more human when I come here."

Tim Robbins and Susan Sarandon are also on the list. They have taken every opportunity to criticize America and alleged that American soldiers had murdered 650,000 Iraqis since the war in Iraq began. About 75,000 have died but most of them were victims of terrorist suicide bombers. Robbins-Sarandon never criticized Saddam Hussein, who murdered hundreds of thousands of his own people.

Next, Danny Glover who once called America "the greatest purveyor of violence in the world." When he visited Fidel Castro in Cuba he called the Bush administration "murderers." Glover, who is also a close friend of Hugo Chavez of Venezuela, once called our president "the devil."

Johnny Depp said no one should want to live in America. Wouldn't it be swell if he were to take his own advice?

Columnist Burt Prelutsky reminds us that back during World War II men like Jimmy Stewart, Tyrone Power, Clark Gable and George Montgomery gave up lucrative studio contracts to serve in the military, quite a contrast to the contemporary leading men in Hollywood.[15]

"Hollywood is such a different place today that it might as well exist on another planet," he said.

Prelutsky says that actor Sean Penn once paid his respects to Saddam Hussein of Iraq.

"You have everyone from Tim Robbins to Billy Crystal slandering the president (Bush) and insisting that America is a far greater threat to democracy than China or Iran," Prelutsky said.

Bruce Willis took his animosity a step further and attacked Christianity when he called it "mythology."

Through the years I've tried to understand why so many movie stars hate America. This has been the land of opportunity for them, the land that opened the doors for fame and unbelievable wealth. The land where millions idolize them and spend billions of dollars to see their movies and purchase their DVDs.

Perhaps when anyone is obsessed with self-love, there isn't much room in his or her heart to love anyone else and that creates a deep vacuum inside the soul.

Self, then, becomes god. But that creates a problem: if they insist on being gods unto themselves, they may fool around and make the real God mad.

The proud and powerful in the entertainment industry often use the medium to revise history to conform to their particular worldview. Warren Beatty's production of

"Reds" is a classic example. "Reds" is a tribute to American journalist John Reed, a communist, who was in Russia during the revolution of 1917 and returned to the United States hoping to lead a similar revolution here.[16]

Now Matt Damon has plans for a new documentary and his attempt to revise American history no doubt will be shown to millions of school children after its run in theaters. Several well-know actors will do readings for the film.

"The actors, who apparently loathe the United States, will be reading from the radical book *A People's History of the United States*," said authors Floyd and Mary Beth Brown in an article entitled "Hollywood's revisionist history."[17]

Radical, left-wing Boston University Professor Howard Zinn wrote the book. His impetus for the work was "to awaken a greater consciousness of class conflict, racial injustice, sexual inequality, and national arrogance," according to the Browns.

"The book is filled with letters, speeches and last words of radicals, revolutionaries and activists, some famous and others unknown," the Browns said.

They point out that the premise of *A People's History* is that America is evil.

Actors who will read for the film include Danny Glover, Viggo Mortensen, Ben Affleck, Kerry Washington, Josh Brolin, Marissa Tomei, Damon and several others.

There is also a definite trend toward atheism in Hollywood.

According to the book *Celebrities in Hell*, by Warren Allen Smith, a number of well-known movie stars and television personalities are atheists, agnostics, humanists and other skeptics.[18]

Through the years a number of Hollywood stars attacked movie icon Charlton Heston for his conservative beliefs including the right to bear arms and the Second

Amendment to the Constitution.

When he passed away in April of 2008, we lost one of our all-time great dramatic actors. He was also a fine Christian gentleman.

I once interviewed him in Tokyo, Japan. He and his wife were in Japan promoting his new movie *The Bible.*

My good friend W. H. "Dub" Jackson, a Southern Baptist missionary, asked Heston to read the Christmas story on nationwide Japanese television and the movie icon agreed.

His wife Lydia told me how wonderful it was for her husband to be doing something for the Lord at Christmastime.

It was a wonderful experience as millions of Japanese sat quietly before their TV sets and listened to Heston – an actor they knew through *The Ten Commandments* and *Ben Hur* – read the Christmas story.

One thing I remember about that interview is that both Charlton and Lydia Heston were kind and gentle people.

Heston was a Christian and faithful in his church attendance; he served as president of the Screen Actors Guild a record six times; served in World War II on a B-25 bomber; marched with Martin Luther King Jr. in Washington; wrote the introduction to Dr. Bernard Nathanson's great pro-life film *The Silent Scream;* was an articulate advocate of public decency and was instrumental in Time-Warner's canceling rapper Ice-T's contract because his songs celebrated the murder of police officers; and served with distinction as president of the American Rifle Association.[19]

When Heston's family announced he had Alzheimer's disease, George Clooney publicly joked about Heston's illness.

"Charlton Heston is the head of the National Rifle Association," Clooney said. "He deserves whatever anyone says about him."

Clooney is a lightweight compared to Heston and should be ashamed of what he said about the great man.

"Making fun of somebody with Alzheimer's disease and feeling no remorse is about as low as it gets, but it isn't all that surprising in this case," said Brian Fitzpatrick writing for *Perspectives*. "To Clooney, Heston's embrace of conservative orthodoxy on the Second Amendment made him worse than... subhuman, not even deserving of the most basic courtesies."

Chapter Eight
Sound and Fury

"Life's but a walking shadow, a poor player
That struts and frets his hour upon the stage
And then is heard no more: it is a tale
Told by an idiot, full of sound and fury,
Signifying nothing..."

Those words by Shakespeare's Macbeth in his famous soliloquy sound like today's agenda-driven media journalists who are all air and hairspray with a whole lot of thunder but not much rain.

These self-proclaimed masters of the media who strut around like a bunch of peacocks are the public relations arm of the secular progressives. In this omnipresent, iconoclastic media, some are so arrogant they have come to believe there are three branches of government: ABC, NBC and CBS. Others believe the media should be the fourth branch of the federal government along with the executive, representative and judicial. Such is their inflated ego.

Extensive polling data among the news media elite is quite revealing. When one pollster asked: "Who do you think should run this country?" they ranked themselves first.

Newspapers and television networks once were magnificent institutions. But somewhere along the way, they went into a tailspin and stumbled and fell into the abyss of moral relativism.

I was the city editor of the afternoon newspaper in

Shreveport, Louisiana, until 1979 and during the last few years as a newspaperman, I became quite disenchanted with my profession. In the 70s I saw the newspapers drift far away from objective, unbiased news coverage, the method most of us learned in journalism school, as reporters began writing their viewpoints or those of editors and publishers into their stories.

Some called it "The New Journalism."

My professors taught me to cover a story by trying to determine the facts then write the story objectively based on the truth. There was no room for my own opinion. They told me the editorial writers would take care of that.

What brought about such a dramatic change in the 70s? Most of the young broadcast and print journalists graduated from liberal universities where socialist professors planted all kinds of radical ideas concerning social change in their young minds.

Little by little, the radical university worldview opened Pandora's Box.

For instance, if a reporter was pro-abortion, his stories began to reflect his bias as he learned to manipulate sources to make sure he quoted only those that supported his view.

Just overnight the media changed and movie stars and radical politicians suddenly were cool but evangelical Christians bigoted; Eastern religions were in, biblical Christianity out; homosexuals and lesbians became a downtrodden minority to be protected; conservatives who disagreed were treated like lepers; young people who became tree huggers and marched in the streets to protect the Spotted Owl were heroes, those who wanted to pray in the schools old-fashioned and out of step with the new morality; when only hundreds marched on Washington, D.C. in favor of abortion, they received front-page coverage and top-story status on the network news, but when thousands marched for life, they were virtually ignored.

Today, in this corrupt media, the progressives wear white hats while Christians and conservatives are treated like Darth Vader.

At the *Shreveport Journal,* our entertainment critic was a homosexual. He wrote a scathing article for the newspaper's weekly magazine about a television minister who often told his viewers what the Bible has to say about the homosexual lifestyle. Fairness would dictate that the writer declare himself to be a homosexual so the readers would clearly understand his bias. But, to my knowledge, that fairness does not exist at any newspaper.

During the last year I served as city editor of the *Journal,* that once-proud newspaper wrote favorable editorials on the new abortion clinic in the city and the pornographic movie theater that had come to town. Also, the publisher instructed the editorial writers to begin a campaign to change a city ordinance that would allow people to drink alcoholic beverages in the city parks on Sunday.

Obviously, I disagreed with those editorials and shared my feelings with the executive editor and the publisher.

Even some of the so-called "family newspapers" often are filled with trash that should not be brought into our homes. They show lewd scenes in their advertisements for X-rated movies and promote escort services compromising once-valued principles for a few more dollars in advertising.

When Shreveport police closed down a pornographic movie house in the city, one of our editorial writers – a far-left secular humanist – soundly chastised the police. His secular progressive bias was clearly seen in the article which had the following headline: "Police Raids Give This Fellow the Blues."

"Unfortunately, I missed this month's weekly raid on the Glenwood and Capri theaters," he wrote. "I wish I had been there… It would have been reassuring to witness the salvation squad's attempts to save me from myself…"

Some journalists like the editorial writer often get mad about the wrong things. He should have been angry about the corrupting influence of pornography in the city, an influence that all surveys reveal leads to rape and other crimes against women and incest in the home. But the writer was only upset with the policemen who were doing their duty. They were acting within the limits of the law which unequivocally says the community has the right to set its own moral standards and the Shreveport community was overwhelmingly opposed to pornography. But apparently the owner and editorial writers at our newspaper were determined to break down all moral restraints in the city.

Now let me show you the methodology a newspaper or television reporter uses to slant public opinion favorably toward a woman's right to choose to terminate a pregnancy, a smut dealer's right to open a porno house, or the rights of men and women to drink beer in the public parks on Sundays.

They ask questions during an interview that promote their bias; use selective quotes and sound bites that fit their agenda; leave out quotes with which the writer disagrees; take quotes out of context to pervert the meaning; place agenda-driven articles or sound bites on the front page or at the top of the news hour; and bury any statements for the opposing point of view on the back pages. And, if there are pictures or video, they make the opposition look like fools and the supporters like stars.

Is this fair? Of course not. Most Americans have a free-born sense of right and wrong. Perhaps that is why all three of the major television networks have lost millions of viewers and most major newspapers large numbers of readers. People are reacting to this "anything goes" journalism and the bias and distortions in these once-great broadcast and print institutions that have prostituted the freedom of the press into a freedom for license.

U. S. News and World Report published an article entitled "Uneasy Press Sets Out to Refurbish Its Image." In the story the news-magazine chronicled a number of incidences where newspaper reporters and writers were guilty of telling half-truths, fabricating stories and writing lies about people."

The magazine said, "The nation's journalists, who for generations have been examining other people's foibles, now find themselves under the public's microscope...

"Exponents of new journalism use composite characters, report events and dialogue that they neither saw nor heard and recount the innermost thoughts of subjects in a combination that can blur the line between fact and fiction."

Hence, I ask the question: Can we trust the news media?

Jayson Blair, 27, a staff reporter for the *New York Times,* was one of the most successful of all journalists at deceiving the public and even his own editors. He wrote stories with datelines from Maryland and as far away as Texas yet never left New York City. An investigation of his work discovered widespread fabrication and plagiarism in his articles that was a profound betrayal of trust both to his newspaper and the American people.

"He fabricated comments," *The Times* reported. "He concocted scenes. He lifted material from other newspapers and wire services. He selected details from photographs to create the impression he had been somewhere or seen someone, when he had not.

"And he used these techniques to write falsely about emotionally charged moments in recent history, from the deadly sniper attacks in suburban Washington to the anguish of families grieving for loved ones in Iraq."[1]

Janet Cooke, a reporter for the *Washington Post,* was nominated for a Pulitzer Prize for her story entitled "Jimmy's World," about a little eight-year-old boy who was a heroin addict in Washington, D. C.

The Pulitzer committee withdrew the feature-writing

prize from Cooke when she admitted the award-winning story was a fraud.[2]

Her story was based on reported interviews with Jimmy, his mother and her boyfriend. However, the reporter admitted she never met or interviewed any of them and that she made up the story after a study of heroin addiction in the city and in conversation with various social workers.[3]

Cooke also gave false personal information to the Pulitzer committee when she said she was a *magna cum laude* graduate of Vassar College. They later learned she only attended the prestigious school for one year.

An associate editor of the *Portland Oregonian* admitted he made up quotes attributed to former Governor Dixy Lee Ray of the State of Washington. He was suspended for sixty days.

When a nut-case went to an Episcopalian church in Seattle with explosives strapped to his body, a columnist for the *Seatle Post-Intelligencer* said she could understand him wanting to blow up a church.

What planet is she living on?

Another legacy from the "new journalism" of the 70s is that the media began equating intellectualism with truth and the result was that nihilism, the idea that traditional beliefs are unfounded, spread through the journalism community like wildfire. Hence, if you believe in abortion, open pornography, homosexual marriage, gratuitous profanity on television and nudity in the movies you are an enlightened intellectual. However, if you believe that those social anomalies are dangerous and destructive influences in our culture, you are considered an ignorant backwoods fundamentalist.

Columnist Prelutsky believes the liberal bias is a reason why so many newspapers are losing circulation numbers.

"People on the left, even those who believe in Santa Claus and the Easter Bunny, deny that liberal bias even exists," he said. "That's because they believe everything they read in the *New York Times* and the *Washington Post*, everything they see on CNN and MSNBC, and every word uttered by Chris Matthews and Bill Maher."

Men and women like Matthews and Maher believe that if they scream loud enough like children, they will get what they want. But they have an innate fear of any opposing point of view. Their agenda is to smear first, last and always those of us who believe in traditional morality.

Prelutsky says that Christians and conservatives tire of seeing everything they believe in being ridiculed by newspapers, not only in the editorial pages but all throughout the paper.

"The question that comes to mind is why a business, any business, would go out of its way to antagonize, depending on the city, between, say, 40 and 60 percent of its market," he said. "Is it possible that while my back was turned, the DNC (Democratic National Committee) bought up every paper in the country except for the *Wall Street Journal* and the *Washington Times?*"

Prelutsky pointed out that the *Los Angeles Times* has lost three hundred thousand in circulation in recent years and said he has personal knowledge that many of those dropped the paper because of what he calls "the paper's relentless bashing of conservatives."

State Senator Dan Richey of Louisiana made some interesting comments about the news media. Speaking out on how the media reported a certain speech by President Ronald Reagan, he said, "The president defined the problem – government overspending – and then mapped out additional ways to reduce our dependence on the federal government...

"If you had the misfortune of watching the president's speech on CBS, you noticed the news reporters tear into the president's program like sharks. They sent out so many negative vibrations in five minutes it's a wonder our TV sets didn't explode...

"CBS and other representatives of the national news media have made a mockery of reporting the news. Their tactics are nothing short of 'yellow journalism,' a phrase coined several generations ago to describe news distortions and exaggerations...

"This may end up having a great positive impact on national politics... Because the sooner the people realize just how biased the national media really is, the less likely they are to be influenced by such media misfits."

Senator Richey is right when he says those in the news media – even network anchors – are like a bunch of sharks.

Remember Dan Quayle? He was vice president under President George Herbert Walker Bush from 1989 to 1993 and, although he was one of the most decent men ever to serve in public office, he was greatly maligned by the malevolent media.

There was a time when I watched the *Today* program nearly every morning. On one of those mornings Katie Couric interviewed Quayle and was disrespectful and mean-spirited, even cruel, toward him. But, as always, he was a perfect gentleman although one could tell by his countenance that he was hurt by the way he was being treated.

I vowed that morning never to watch Couric or the *Today* program again. Now, some twenty-five years later, I continue my personal boycott of NBC and Couric, even though she has moved over to CBS and has some of the lowest ratings in CBS new history.

In July of 2007, Couric, breathing fire, slapped an editor at CBS because of a certain news story. And she, with her smug self-righteousness, wants to tell Vice President Quayle and the rest of us in America how to run our lives? I don't think so.

I have a theory that some newsmen and women have inner demons that range from deeply inflated opinions of themselves to painful self-doubt. I've wondered if these pathologies lead them to believe they can build themselves up by tearing some famous person down.

President Ronald Reagan was a master at handling those self-aggrandizing reporters. When they would purposely try to irritate him, he would smile, shake his head and say, "Well, there you go again" as he refused to dignify their questions with answers.

Certain members of the news media can also be masters of deceit.

When I was a free-lance war correspondent in Vietnam in 1967, I was amazed how some network reporters filed their stories on the war. I saw them day after day, dressed in tiger uniforms with bush hats pinned up on one side like Australian jungle fighters, report from the garden of the Intercontinental Hotel in Saigon which gave the impression to viewers they were in the jungles with our troops.

Ironically, while some of us were on jungle patrols with various units of the First Air Cavalry Division or on search-and-destroy missions, those network reporters were enjoying the air conditioned rooms at the Intercontinental and American beer in the bar. Some of them even hired Filipino and Korean cameramen to handle all the dangerous assignments. I did meet several journalists on the front lines including a *Newsweek* reporter and one from a newspaper in Indiana and I'm sure there were many others.

It's no wonder some of the networks turned against our troops. They didn't have a clue as to what was really going on during the war because they were getting their reports from the garden at the Intercontinental Hotel.

I visited a Marine Corps hospital in Da Nang that handled neurological wounds and I saw every kind of head wound imaginable. As I looked at those young men, I thought, "These are America's best. How can we ever justify our men suffering like this in a foreign land facing a fierce and dedicated enemy and being hated by the news media and so many other folks back home?"

After leaving Vietnam I had some misgivings about the war. Not about our soldiers but the politicians in Washington who couldn't make up their minds whether to win the war. I loved the troops who were doing what a civilian government called on them to do. They deserved my honor, respect and support.

Our nation's courageous attempt to preserve the freedom of the people of South Vietnam has had some positive influence in Southeast Asia and around the world. For instance, since that war there has been no other country taken over by the communists who realized that Americans will fight for other peoples' freedom even at the cost of the lives of our best young men and women.

Finally, when I returned to the states I was amazed at the anti-war sentiment in the network news reports. They inferred that a large number of our men in Vietnam were using drugs, refusing to fight the enemy, fragging their officers with hand grenades, and were on the verge of mutiny. But the men I met in Vietnam were doing their very best to defend the people of South Vietnam and win the war. I saw burly medics going out into the villages and taking care of little Vietnamese children, giving them inoculations, and treating their wounds. I never saw those acts of kindness reported on the six-o'clock news.

The news media, filled with sound and fury, had what I call a "bee-hive mentality" during the war. When the *New York Times* took a position on the war the other media – *The Boston Globe, Washington Post, ABC ,CBS, NBC* – joined the parade. That kind of reporting cuts deep into the respected concept of objectivity and fairness in journalism.

We supposedly live in a pluralistic society. Yet the media often seek to segregate those of us with conservative, Christian values out of that pluralism and we get pushed around by the multi-billion dollar corporate news giants.

One year, the internet giant Google celebrated the

achievement of the communist Sputnik on the fiftieth anniversary of the historic flight into space but routinely ignores American patriotic celebrations such as Memorial Day and Veteran's Day.

WorldNetDaily noted that for nine years Google declined to commemorate Memorial Day but has modified its logo to celebrate the Chinese New Year, Valentine's Day, Halloween and several other observances.

Google has also celebrated National Teachers Day, Women's Day, Ray Charles' birthday, World Water Day and St. George's Day.

"Besides overlooking Veterans Day and Memorial Day since the company's inception in 1999, it also has ignored Christmas," WorldNetDaily reported.

"What would (King) David say if he were alive today and witnessed the propaganda and promotion that make up what we casually call 'the media'?" asks bestselling author Warren Wiersbe. "He would probably describe today's 'communication' as he did centuries ago: empty and useless words ('vanity'), smooth talk ('flattery'), double-talk from double hearts…"[4]

Wiersbe thinks it is paradoxical that "society today sees the Scriptures as something relatively worthless and yet pays great sums of money to the people who manufacture deception…"

"If God's people will saturate themselves with God's Word, they won't be seduced by 'this lying generation,'" Wiersbe said.

He points out the media exalts those things God condemns: "immorality, brutality, murder, lies, drunkenness, nudity, the love of money, the abuse of authority."

"The things that God condemns are now a means of universal entertainment, and the entertainment industry gives awards to the people who produce these things," Wiersbe said.

Chapter Nine
A Lot of Hot Air?

Here are some very interesting newspaper headlines from 2009:

"Midwest Bracing for Heavy Snow – Wind Chills of 50 Below."

"Next Arctic Blast Blows Even Colder."

"Britain's Big Snow Shuts Cities."

"Arctic Blast Freezes Texas."

"Weekend Freeze Looms for Gulf Coast."

"Florida Races to Save Crops."

Global warming? Really?

In the midst of all the panic and hysteria over global warming, several new studies reveal it is just a lot of hot air.

More than 31 thousand scientists throughout the United States have signed a document known as the Petition Project rejecting the assumption that man through greenhouse gases is causing global warming and harming the Earth's climate.[1]

This is the largest group of scientists to date to speak out on the global warming controversy. Among the scientists there are more than 9,000 Ph.D.s in the fields of atmospheric science, climatology, Earth science, environment and several other scientific disciplines.

"There is no convincing scientific evidence that human release of carbon dioxide, methane, or other greenhouse gases is causing or will, in the foreseeable future, cause catastrophic heating of the Earth's atmosphere and disruption of the Earth's climate," the petition states. "Moreover, there is substantial scientific evidence that increases in atmospheric carbon dioxide produce many beneficial effects upon the natural plant and animal environments of the Earth."

The project was launched back in 1998 when several thousand scientists first signed the petition.

"Then, between 1999 and 2007, the list of signatures grew gradually without any special effort or campaign," said Bob Unruh writing for *WorldNetDaily*.

However, recently there has been a concerted effort to enlist scientists to sign the petition because of former Vice President Al Gore's "escalation of the claims of 'consensus'" and the release of the movie *An Inconvenient Truth*' by Gore and certain other events, according to officials with the Petition Project.[2]

"Mr. Gore's movie, asserting a 'consensus' and 'settled science' in agreement about human-caused global warming, conveyed the claims about human-caused global warming to ordinary movie goers and to public school children, to whom the film was widely distributed," Art Robinson, a spokesman for the project said. "Unfortunately, Mr. Gore's movie contains many very serious incorrect claims which no informed, honest scientist could endorse."[3]

However, the film is being shown to students in many high schools and colleges throughout the land.

Robinson pointed out that recently Gore said global warming causes cyclones, typhoons and even shark attacks. He attributed the cyclone that devastated Myanmar in the spring of 2008 to global warming.

During an interview with National Public Radio, Gore also referred to a "catastrophic storm" that hit Bangladesh and "the strongest storm in more than 50 years hit China – and we're seeing consequences that scientists have long predicted might be associated with continued global warming."

And now each time the sun goes behind a cloud, someone in the news media yells, "Yeah, global warming!"

But, as usual, Gore is way out in left field.

Professor Fredrick Seitz, the past president of the U. S. National Academy of Sciences and winner of the National Medal of Science, before he passed away wrote a letter promoting the petition and opposing an international treaty that would restrict the use of technologies that depend on coal, oil and natural gas.[4]

"This treaty is, in our opinion, based upon flawed ideas," Seitz said. "Research data on climate change do not show that human use of hydrocarbons is harmful."

"A major new scientific study concludes the impact of carbon dioxide emissions on worldwide temperatures is largely irrelevant..." says Dr. Reid Bryson, founding chairman of the Department of Meteorology at the University of Wisconsin who added that the temperature of the earth is increasing, but that it has nothing to do with carbon dioxide emissions or greenhouse gases.[5]

Dr. Bryson acknowledges temperatures are going up. "It has gone up since the early 1800s, before the Industrial Revolution, because we're coming out of the Little Ice Age, not because we're putting more carbon dioxide into the air."

But he refers to a scientific project entitled "Heat Capacity, Time Constant, and Sensitivity of Earth's Climate System," by Stephen Schwartz of the Brookhaven National Laboratory which casts doubts concerning catastrophic, man-made global warming.

Some scientists believe the United States will waste trillions of dollars trying to solve the so-called global warming problem which is no problem at all.

Gore, the high priest of global warming, won an Oscar for *An Inconvenient Truth* and also received the Nobel Peace Prize for his work.

I conducted my own research on some of the world's worst disasters and the result is quite interesting.

For instance, according to Wikipedia Encyclopedia, there was an earthquake in Egypt in 1201 that killed 1.1 million people and another in China in 1556 that killed 830,000. There was a cyclone in India in 1839 that killed 300,000 and another in Bangladesh in 1970 that killed 500,000. These are far worse than the disaster in Myanmar yet no one could possibly attribute those disasters to global warming.

Any member of the news media could have discovered the same information. However, secular progressive members of the news media and scientists are not interested in any information that refutes global warming. But it is quite evident that the disasters in China, India and Bangladesh and hundreds of others took place before the so-called "crisis" in greenhouse gases that Gore predicts will soon destroy the earth.

Sometimes I wonder why the left-wing media and secular progressive scientists won't listen to the truth.

For instance, physicist Phil Chapman, the first Australian to become an astronaut with NASA, noted there was a cooling period in the world of about 0.7 degrees Centigrade between January of 2007 and January of 2008.

"This is the fastest temperature change in the instrumental record, and it puts us back to where we were in 1930," Chapman wrote in *The Australian* in 2008.

"My guess is that the odds are now at least 50-50 that we will see significant cooling rather than warming in coming decades," Chapman said.

Why do the progressives in the media and academia continue to consider global warming an absolute and refuse to even consider the possibility that Gore and the alarmist scientists could be wrong? It's simple. The idea of global warming gives the progressives an open door to criticize our present form of government, big business including the automobile industry, and capitalism and free enterprise which they believe spawn the worst offenders contributing to the warming. Their solution to the problem is government control by a government that is already out of control. They know that in order to usher in an age of socialism – or complete government control – in this land, all the institutions that made this country great must be torn down. Global warming is a step in that direction.

Have you heard the news about Antarctica? It has ninety-two percent of the earth's ice and is getting colder every year and the icecap is growing larger every year.

But that's not all. The U. S. National Oceanic and Atmospheric Administration (NOAA) reported in February of 2008 that most of the so-called "lost" ice in the Arctic – that supposedly disappeared because of global warming – is almost back to its original level, according to Phil Brennan, a veteran journalist who writes for *NewsMax*.[6]

The NOAA report acknowledged that the earth's ice shrunk from five million square miles in January of 2007 to one-and-a-half million square miles in October. But today there is nearly one-third more ice in Antarctica than in years past.

According to the *London Daily Express*, the growing icecap in Antarctica has raised questions for the crusaders for global warming such as Gore and has encouraged the skeptics who do not believe the earth is facing a global warming catastrophe.[7]

"The *Daily Express* recalls the photograph of polar bears clinging on to a melting iceberg which has been widely hailed as proof of the need to fight climate change and has been used by... Gore during his 'Inconvenient Truth' lectures about mankind's alleged impact on the global climate," Brennan said. "Gore fails to mention that the photograph was taken in the month of August when melting is normal. Or that the polar bear population has soared in recent years."

According to Brennan, scientists acknowledge that in 2007 the northern Hemisphere experienced the coldest winter in several decades and the heaviest snow since 1966.

"If global warming gets any worse we'll all freeze to death," Brennan said.

However, one survey by a media watchdog group found that the major newspapers and television networks have written or broadcast twenty-seven stories on warming in the Arctic – and the commensurate shrinking of the icecap – to one story on the increasing colder temperatures and expansion of the icecap in Antarctica.

Does that surprise you?

In March of 2008, National Public Radio broadcast a segment titled "The Mystery of Global Warming's Missing Heat."

"Some 3,000 scientific robots that are plying the ocean have sent home a puzzling message," NPR's Richard Harris reported.[8] "These diving instruments suggest that the oceans have not warmed up at all over the past four or five years."

Scientists believe that eight percent to 90 percent of global warming involves warming ocean waters.

Willis has been studying the ocean depths with a fleet of robotic instruments called the "Argo System," NPR's Harris reported.

"The buoys can dive 3,000 feet down and measure ocean temperature," Willis said. "Since the system was fully deployed in 2003, it has recorded no warming of the global oceans."

Actually, Willis says he has discovered "a slight cooling" in the oceans.

Willis, who apparently believes that global warming is a threat to the earth, thinks that there is some scientific answer as to why the oceans are not warming, but cooling and scientists will solve the puzzle in the future.

But Gore, like Cervantes' Quixote of old who valiantly fought the windmills he thought to be giants, continues to make fun of anyone who disagrees with his position.

Early in April of 2008, Gore appeared on the popular newsmagazine program *60 Minutes* where he was interviewed by veteran newswoman Leslie Stahl.

Stahl asked Gore what he thought about then-Vice President Dick Cheney and others who question whether humans are the cause of global warming.[9]

"I think that those people are in such a tiny minority... now with their point of view, they're almost like the ones who still believe that the moon landing was staged in a movie lot in Arizona, and those who believe the earth is flat," Gore replied.

However, Lord Christopher Monckton, former policy advisor to British Prime Minister Margaret Thatcher during the 1980s, says that for the present time Gore can enjoy his jokes about "flat earth fantasies, but that in years to come people will laugh at him.

"The alarmists are alarmed, the panic mongers are panicking, the scare mongers are scared..." Monckton said. "Why? Because global warming stopped ten years ago; it hasn't got warmer since 1998. And in fact in the last seven years, there has been a downturn in global temperatures equivalent on average to about [or] very close to one degree Fahrenheit per decade. We're actually in a period... of global cooling."

According to Monckton, Gore is in a panic mode for several reasons.

He says Gore has staked his reputation on "telling the world that we're all doomed unless we shut down 90 percent of the Western economies."

Monckton points out that Gore's organization The Alliance for Climate Protection is planning to launch a $300 million advertising campaign to effect reforms to reduce the so-called "climate crisis."

"But Monckton points out that in the U. K. (United Kingdom), Gore is not allowed to speak in public about his 'green investment company' because to do so would violate racketeering laws by 'peddling a false prospectus,'" said Chad Groening writing for OneNewsNow.com. "He says that fact came about after a British high court found Gore's movie, *An Inconvenient Truth,* riddled with errors."

Although Gore maintains we're killing our planet with greenhouse gases and we must make critical changes to avoid a worldwide catastrophe, he doesn't even try to explain what is happening in Antarctica! As a matter of fact, he doesn't even mention it.

According to his movie:
- The people of the world are sitting on a ticking time bomb.
- We have only ten years before apocalypse.
- There will be floods, droughts, epidemics and killer heat waves throughout the world.[10]

"With wit, smarts and hope, An Inconvenient Truth ultimately brings home Gore's persuasive argument that we can no longer afford to view global warming as a political issue – rather, it is the biggest moral challenge facing our global civilization," the promotion trailer to the movie asserts.

Hence, something radical must be done to save the planet before it burns up and destroys us all.

Ted Turner, environmentalist and founder of the Cable News Network, agrees with Gore and says that by mid-century the earth will be destroyed.

Turner says that unless drastic steps are taken, "We'll be eight degrees hotter in 30 or 40 years and basically none of the crops will grow." He made the comment during an interview with broadcast journalist Charlie Rose on the PBS.

However, Turner has his own solution – population

control.

"We're too many people; that's why we have global warming," he said.[11]

Turner suggested that every family in the world pledge not to have more than one or two children in order to cut down on the world's population.

But what is the real truth about global warming?

Hundreds of scientists believe Gore's argument is a lot of hot air.

Could it be that the alarmist theory is based on ideology rather than science and therefore is agenda-driven? Is it an ideology promoted through fear-mongering tactics of environmentalists, the media and secular progressives, an ideology that is against big business, free enterprise and capitalism?[12]

"Have you noticed, for example, how every heat wave results in more media hysteria, while record cold waves are brushed aside as irrelevant?" asks Fred L. Smith, Jr. "And remember the dire predictions that the record 2005 hurricane season was a trend that would continue because it was fueled by global warming? Remember how it was followed by an embarrassed media in 2006, when not a single hurricane hit the U.S.?"

Smith is president and founder of the Competitive Enterprise Institute, a free market public policy group. He is a frequent guest on CNN's "Crossfire," PBS' "News Hour with Jim Lehrer" and "Now" with Bill Moyers, "20/20" and "This Week" and other programs.

Smith says the alarmist environmentalists never give up. Gore and his followers have enlisted famous Hollywood stars and other pop culture icons and the global warming movement is on the verge of catapulting the agenda into government policy that will cost trillions of dollars and cripple the automobile industry and coal-fired power plants.

The apparently unstoppable green political juggernaut has also convinced the secular progressives in Congress that this planet will self-destruct unless they take action. And they control the nation's money.

Smith points to two scholarly books that challenge the claims of global warming. One is *Unstoppable Global Warming Every 1,500 Years* by scientists Fred Singer and Dennis Avery. The authors present scientific evidence that there have been 600 global warming and cooling periods in the last million years. The second is *The Chilling Stars* by Professor Henrik Svensmark and Nigel Calder. The book explains how cosmic rays amplify even minute changes in the brightness of the sun, causing a one-to-two degree temperature increase on earth.[13]

"Global warming alarmists would have you believe that, until now, the Earth has always had a stable climate," Smith said. "But in fact, over the millennia, the earth has warmed and cooled repeatedly."

Gore and his followers and the green lobby believe there is absolute scientific proof to support an imminent climate disaster. He once said "the debate in the scientific community is over."

But if Gore and his followers are so sure of the scientific evidence for the warming, why is there such a strong reaction when anyone questions their views? Smith wonders.

"I believe the answer is that they're not all that certain..." Smith said. "They know full well that many of their claims can't withstand critical review. And they know that once their pretense of a consensus starts falling apart, their radical political agenda will collapse as well."

John Coleman, meteorologist and founder of The Weather Channel, says the manmade global warming myth is the "greatest scam in history."

"I am amazed, appalled and highly offended by it... Some dastardly scientists with environmental and political motives manipulated long term scientific data to create an allusion of rapid global warming," he wrote in an article for ICECAP.

Coleman pointed out that a number of other scientists jumped on the band wagon to support and broaden the "research" to further enhance "the totally slanted, bogus global warming claims."

"Their friends in government steered huge research grants their way to keep the movement going," he said. "Soon they claimed to be a consensus."

According to Coleman, the scam includes environmental extremists, powerful politicians, movie, media and other environmentalist journalists. He wrote that they have created a wild "scientific" scenario warning the public of dire environmental consequences from global warming unless everyone adheres to their radical agenda.

"Now their ridiculous manipulated science has been accepted as fact and become a cornerstone issue for CNN, CBS, NBC, the Democratic Political Party, the Governor of California, school teachers and, in many cases, well informed but very gullible environmental conscientious citizens," he wrote.

Coleman, who has discussed so-called global warming with numerous scientists, said.

"There is no run-away climate change. The impact of humans on climate is not catastrophic. Our planet is not in peril. I am incensed by the incredible media glamour, the politically correct silliness and rude dismissal of counter arguments...

"In time, a decade or two, the outrageous scam will be obvious."

Some California lawmakers believe the schools in the state should teach reading, writing and global warming.

State Senator Joe Simitian of Palo Alto has introduced legislation to require the science curriculum to cover climate change.[14]

The bill, according to Simitian, would require that all future science textbooks approved for use in the state include information on climate change.

"You can't have a science curriculum that is relevant and current if it doesn't deal with the science behind climate change," Simitian said. "This is a phenomenon of global importance and our kids ought to understand the science behind that phenomenon."

However, some of the other senators are suspicious of the proposed legislation.

"I find it disturbing that this mandate to teach this theory is not accompanied by a requirement that the discussion be science-based and include a critical analysis of all sides of the subject," said Sen. Tom McClintock of Thousand Oaks during the Senate debate.[15]

Sen. Jeff Denham of Modesto said he wants guarantees that the views of those who do not believe in global warming also be presented to schoolchildren.

"We don't have complete factual information yet," Denham said. "From what I have seen the Earth has heated and cooled on its own for centuries."

Professor emeritus Reid Bryson, known as the father of scientific climatology, does not believe in man-made global warming. Neither does the retired University of Wisconsin-Madison professor subscribe to the scientific consensus on the matter.

"But he is not skeptical that global warming exists, he is just doubtful that humans are the cause of it," Samara Derby wrote in *The Capital Times*. "There is no question the earth has been warming. It is coming out of the 'Little Ice Age.'"

Professor Bryson says there is no solid evidence that the warming is caused by the people on earth or by carbon dioxide. He agrees that humans are polluting the air and sending a lot of carbon dioxide into the atmosphere but the effect on the earth's temperature is minor.

He said that the fact that most scientists believe in global warming proves absolutely nothing. "Consensus doesn't prove anything, in science or anywhere else…"

Bryson was the founding chairman of the department of meteorology at the university and of the Institute for Environmental Studies.

Meanwhile, one of the world's best-known meteorologists has called Gore's theory of Global Warming "ridiculous" and the result of "people who don't understand how the atmosphere works."[16]

Dr. William Gray made his comments during a lecture to the students at the University of North Carolina. He told the students that humans are not responsible for the climate change in the earth.

"We're brainwashing our children," said Dr. Gray, a tenured professor at Colorado State University. "They're going to the Gore movie 'An Inconvenient Truth' and … it's ridiculous."

"During his speech to a crowd of about 300 that included meteorology students and a host of professional meteorologists, Dr. Gray also said those who had linked global warming to the increased number of hurricanes in recent years were in error," wrote Steve Lytte in *The Sydney Morning Herald.*

He presented statistics that reveal there were more hurricanes during a period of cooler global temperatures than from 1957 to 2006 when the earth was warmer.

"The human impact on the atmosphere is simply too small to have a major effect on global temperatures," he said.

Wesley Pruden, the former editor of *The Washington Times,* said that a number of scholars are speaking out concerning climate change.

"Richard Lindzen, the Alfred P. Sloan Professor of Meteorology at the Massachusetts Institute of Technology, scoffed at the wishful claims that the debate (on global warming) is over," Pruden wrote in his column.

Lindzen said in an interview with *Newsweek* that there are those who believe the earth is facing a serious crisis that requires immediate action. But he pointed out there is "no compelling evidence that the warming trend we've seen will amount to anything close to catastrophe."[17]

"Indeed, the only catastrophe anyone can see this week is the brutal... snowstorm in the Midwest, breaking records for cold and ice in Chicago, Cleveland and even North Dakota," Pruden said. "The peaches, apples, blueberries and grapes in a wide swath of the Southeast from Arkansas eastward to the Carolinas, were ruined by record-breaking freezes."

He pointed out that global temperatures are averages compiled over many years "but the spring freeze was a reminder that when man proposes God disposes, whether we believe in him or not."

Pruden likes to poke fun at the frenzied global warming alarmists including certain entertainers.

"Miss (Sheryl) Crow is a pop singer of some repute, as such reputations go, but wants to do bigger things than sing songs..." Pruden said. "She's busy at the moment trying to cool down the globe, having just completed a transcontinental bus tour with her gal pal... promising bemused college kids there's soon going to be a hot time in the old town, if not tonight then just as soon as the April ice and snow melt."

Pruden quoted the following from her blog: "Although my ideas are in the earliest stages of development, they are, in my mind, worth investigating," she said. "One of my favorites is in the area of forest conservation which we heavily rely on for oxygen. I propose a limitation be put on how many squares of toilet paper can be used in any one sitting… I think we…can make it work with only one square per restroom visit…"

According to Pruden, Crow travels the country with an entourage of three tractor trailers carrying sound and other equipment, four buses and six cars "to spread her message of frugality to others."

"Her typical concert contract includes binding instructions that she must have in her dressing room 12 bottles of Grolsch beer, six bottles of 'local beer,' …bottles (one each) of 'good' Australian cabernet, a merlot, bourbon, gin and brandy…and lots of chips and dips"

Pruden adds: "On a diet like that, one square of toilet paper is never enough."

He also noted that, according to Gore and his crowd, all of us should already be cooked medium-rare by global warming.

"Global warming is scheduled to kill us all before next Christmas," Pruden said in November of 2007.

But he explained that since Christmas has gone the way of the hula hoop to keep from offending Osama bin Laden, the ACLU and various assorted grinches, we all may be spared from the prophesied catastrophe.

NBC Television gave Gore the biggest in-kind political contribution in history for the three-hour infomercial on global warming. However, the network lost out in the ratings to "Cops" and "America's Funniest Home Videos."

Gore called the three-hour prime time program "the largest global entertainment even in all of human history" and he received 75 hours of free airtime on NBC, CNBC, Bravo, the Sundance channel, Universal HD and Telemundo.[18]

There are some conservative commentators who see a secret agenda behind the world's obsession with global warming.

"The U. N. recently announced global warming is heading inexorably to global catastrophe..." *Whistleblower Magazine* reported.[19] "The news media beat the drum of 'climate catastrophe' daily... And across America school children are frightened to death with tales of rising oceans, monster tornadoes, droughts and millions dying – all because of man-made global warming."

However, U. S. Senator James Inhofe stated that "for more than 100 years, journalists have quoted scientists predicting the destruction of civilization by, in alternation, either runaway heat or a new Ice Age."

The major media in this land have predicted global catastrophe four different times and, each time, speculated that billions of people would die because of a lower yield of crops for food production and that entire countries would be destroyed.

"In 1895, the panic was over an imminent ice age," *Whistleblower* reported. "Later, in the late 1920s, when the earth's surface warmed less than half a degree, the media jumped on a new threat...which continued into the late 1950s."

However, in 1975 the *New York Times* published a story titled "A Major Cooling Widely Considered to Be Inevitable." Six years later, in 1981, the news media reports returned to global warming and the *Times* wrote another story based on interviews with seven government atmospheric scientists who prophesied warming of an "almost unprecedented magnitude."

According to *Whistleblower*, some people including politicians, scientists and big corporations have a lot to gain by frightening billions of people by using the unproven theory of man-made warming.

A panel of eighteen scientists from eleven countries has reported to the U.N. that the only way to stop the climate change and save the world is through a tax on all the countries of the world, a tax on greenhouse emissions, *Whis-*

tleblower reported.

"With communism largely discredited today – after all, 100-150 million people died at the hands of communist 'visionaries' during the last century – elitists who desire to rule other people's lives have gravitated to an even more powerful ideology," *Whistleblower* explained. "More powerful because it seems to trump all other considerations, as it claims the very survival of life on earth is dependent on implementing its agenda."

Hence, the rush toward global governance, with unprecedented worldwide taxation, is being planned for us all.

Actually, climate change of global warming is nothing new.

As far back as 1922, the *Washington Post* carried an article titled "Arctic Ocean Getting Warm; Seals Vanish and Icebergs Melt."

According to the article, during that time period entire glaciers disappeared.

Scientists at NASA estimate that four of the hottest 10 years in the United States were back in the 1930s and 1934 was the warmest.[20]

Apparently the people in Siberia haven't heard about global warming.

According to a report from the AHN Worldwide News Network, government officials in 2007 warned the people of Siberia to be prepared for the coldest winter in history with temperatures reaching sixty-seven degrees below zero.

Although the majority of scientists, news media and congressmen have embraced the concept of global warming, more than 500 scientists have published articles questioning the man-made global warming hysteria.[21]

According to a report from the Hudson Institute, more than three hundred scientists have found evidence that there has been a moderate 1500-year climate cycle that has caused a dozen or more incidences of global warming since the Ice Age and that current warming is related to the sun's irradiance.

"This data and the list of scientists make a mockery of recent claims that a scientific consensus blames humans at the primary cause of global temperature increases since 1850," said Hudson Institute Senior Fellow Dennis Avery.

Although the scientists who question the cause of global warming have written articles for prestigious journals such as *Science; Nature;* and *Geophysical Review Letters,* their findings have seldom been reported by the media.

The Hudson Institute is a non-partisan policy research organization.

Canadian scientists believe that the sun has a greater impact on the climate of the Earth than all the automobile tailpipes and smoke stacks in the world.[22] And some Canadian scientists are more concerned about global cooling than warming. They have recorded data that shows solar activity fluctuates in eleven-year cycles.

"But so far in this cycle, the sun has been disturbingly quiet," the *Investor's Business Daily* reported.

Kenneth Tapping, a solar researcher with Canada's National Research Council, believes that if the sun remains quiet for another year or two, it could precipitate a time of severe cooling of the Earth, with massive snowfall and severe cold weather in the Northern Hemisphere.

"As we have noted many times, perhaps the biggest impact on the Earth's climate over time has been the sun," *Investor's Daily* reported.

After 45 inches of snow fell in New Hampshire in one winter month, *Boston Globe* columnist Jeff Jacoby asked: "Where did global warming go?"

Then there is the so-called "Climategate" where erstwhile reputable scientists covered up scientific data that raised questions about global warming.

Syndicated columnist Thomas Sowell wrote a column in December of 2009 where he said, "Like anything valuable, science has been seized upon by politicians and ideologues, and used to forward their own agendas."

"This started long ago, as far back as the 18th century, when the Marquis de Ciondorcet coined the term 'social science' to describe various theories he favored," Sowell said. "In the 19th century, Karl Marx and Friedrich Engels distinguished their own brand of socialism as 'scientific socialism.' But in the 20th century, all sorts of notions wrapped themselves in the mangle of 'science.'"

According to Sowell, the "global warming" hysteria "is only the latest in this long line of notions, whose main argument is that there is no argument, because it is 'science.'"

"The recently revealed destruction of raw data at the bottom of the global warming hysteria, as well as revelations of attempts to prevent critics of this hysteria from being published in leading journals, suggests that the disinterested search for truth – the hallmark of real science – has taken a back seat to a political crusade," he said.

Michael Barone, a senior writer for *U. S. News and World Report,* says: "Quick, name the most distrusted occupations. Trial lawyers? Pretty skuzzy... Used car dealers? Always near the bottom of the list..."

But, according to Barone, a new group has moved up on the list of mistrusted professions, the climate scientist. "First came the Climategate e-mails made public in November (2009) that showed how top-level climate scientists distorted research, plotted to destroy data and conspired to prevent publication of dissenting views."

Chapter Ten
"Who Left the Door Open?"

Thus began an in-depth story in *Time* magazine concerning run-away illegal immigration into this land, immigration that predictably increased after President George W. Bush proposed an amnesty program for millions of them in 2008. Amnesty would declare them legal and allow them to remain in the United States.

President Bush and Senator Edward Kennedy, now deceased – the odd couple of national politics – joined forces to work on the amnesty bill to make sure that "no illegal alien was left behind."

Various surveys reveal that about 12 million would be affected by amnesty but maybe as many as 20 million. No one knows for sure for the vast majority of them are undocumented.

There are estimates that as many as four thousand cross the border from Mexico each day or three million a year through "the busiest unlawful gateway in the U.S., the 375-mile border between Arizona and Mexico," *Time* reported.

"No searches for weapons. No shoe removal. No photo-ID checks," like those in national airports, the report continues. "Before long, many will obtain phony identification papers, including bogus Social Security numbers, to conceal their true identities and mask their unlawful presence."

According to the *Time* report, some law-enforcement officers believe the mass movement of illegals "offers the perfect cover for terrorists seeking to enter the U.S..."

The report emphasizes that most Americans want to crack down on illegal aliens entering this country, but neither the White House nor the Congress seem to want to deal with the problem.

"When the crowds (of aliens) cross the ranches along and near the border, they discard backpacks, empty Gatorade and water bottles and soiled clothes," the magazine reported. "They turn the land into a vast latrine, leaving behind revolting mounds of personal refuse... Night after night, they cut fences intended to hold in cattle and horses...The immigrants steal vehicles and saddles. They poison dogs to quiet them."

It should be apparent to everyone that something is terribly wrong. Immigration from Mexico and other Latin American countries is a problem of epic proportions.

"In the United States, immigration is in a state of anarchy – not chaos, but anarchy," said James Walsh, formerly with the Immigration and Naturalization Service.

Where does the constitution authorize the president and Congress to grant citizenship to illegal immigrants? Or drivers' licenses?

Polls show that 77 percent of the American people oppose amnesty and eighty-two percent drivers' licenses for illegals.

Yet a host of politicians at the local, state and federal levels are determined to go against the will of the people. For instance, even though our Social Security program is on the fast track to bankruptcy, United States senators voted to give social security benefits to the illegal immigrants.

Somebody needs to say something, loud and clear!

"It's absurd to reward aliens who have no regard for the rule of law by giving them driver's licenses," says James L. Lambert a guest columnist for *OneNewsNow*. "Even some liberals... comprehend this."

Nationally syndicated talk-show host Paul McGuire argues that giving driver's licenses to illegals is "a deceptive way to do an end-run around immigration laws... A driver's license to illegals opens the door to getting things like checking accounts, free government services that U. S. taxpayers must pay for – and it will (potentially) give them the power to vote."

McGuire explains what he believes is the motivation for giving licenses to illegals. "The real purpose… is to mobilize millions of voters who will now vote for big government that will provide them with free services' like food stamps, education and medical care," he said. "Of course, American citizens will have to pay for these services through increased taxes…"

The projected annual cost for taking care of the immigrants in Los Angeles County is $444-million dollars per year.

"This new information shows an alarming increase in the devastating impact illegal immigration continues to have on Los Angeles taxpayers," Los Angeles County Supervisor Michael Antonovich said. "With $220 million for public safety, $400 million for healthcare, and $444 million in welfare allocations, the total cost for illegal immigrants to County taxpayers far exceeds $1 billion a year – not including the millions of dollars for education."

Some observers believe the cost of educating the children of the illegal immigrants could cost another $1 billion each year.

A former economist for the State of California says that taking care of illegal immigrants in the state costs the legal residents tens of millions of dollars each year.

"And he argues that the current federal deficit would be wiped out if illegal aliens were no longer living in the U.S.," said Jim Brown writing for *OneNewsNow*.

Former Governor Eliot Spitzer of New York, who resigned in March of 2008, managed to alienate just about everyone in the state when he arbitrarily decided to give driver's licenses to illegals. There was strong opposition in the state assembly and from the 77 percent of the citizens of New York who opposed the governor's plan.

While the governor was running around the state like Reddy Kilowatt promoting his driver's license scheme, I came across a story about the terrible living conditions in some areas of New York City. Cindy Hsu of CBS News reported that in some poor housing projects the tenants are exposed to mold, mildew, rat infestation, bedbugs and cockroaches.

The thought occurred to me that the governor was spending a lot of his time trying to take care of the illegals when he should be taking care of the poor citizens in his state's largest city.

According to a study by the Census Bureau, if all the aliens presently in the United States were given amnesty – as proposed by President George W. Bush, Sen. Edward Kennedy and a host of others – the yearly cost would jump to $29 billion dollars a year.

One study revealed that each illegal-alien household used $2,700 a year more in services than it paid in taxes. Expenditures included Medicaid ($2.5 billion); medical treatment for the uninsured ($2.2 billion); food assistance programs ($1.9 billion); federal prison and court systems ($1.6 billion): and federal aid to schools ($1.4 billion).

Although illegals do cause a financial problem for the government, those who work pay a substantial amount of federal taxes. But no one knows for sure how many of the illegals have jobs.

One study revealed that illegal immigration is costing the so-called "border counties" along the border between Arizona and Mexico millions of dollars each year just for the prosecution of illegal immigrants for crimes and misdemeanors.[1]

"The study says the battle over illegal immigration is also diverting money from parks, libraries and other law-enforcement efforts," the Associated Press reported.

The question of illegal immigration is a burning issue in Northern Virginia as well as the border states along the Rio Grande River and in the 2007 elections the voters were seeking candidates who promised to deal decisively with the issue.

Voters there complain that illegal immigrants have degraded the quality of life, caused overcrowding in housing, schools, hospitals and jails. They also attribute the increase in gang violence to the influx of immigrants into the area.

To deal with the situation in Prince William County, officials are checking the status of all illegal immigrants who are detained or arrested by police and have cut off services to the illegals.

Dr. Philip Romero, a professor at the University of Oregon, said that for each dollar the illegal immigrants pay in taxes, they receive $8 to $12 in free government services.[2]

He says this "underground" population consumes about 20 percent of the entire state budget. So as a result, taxpayers are paying millions for healthcare, education and the criminal justice system.

A California school teacher who wants to remain anonymous explained how taxpayer funds are being used, and often wasted, in the school systems in the state. She is in charge of the English-As-A-Second-Language Department in a large school in Southern California and most of her students come from poor illegal immigrant families.

She shared the story with the Center for U.S.-Mexico Immigration Analysis.

The school provides free breakfasts and lunches for the children but, she said, "The waste of the food is monumental, with trays and trays of it being dumped in the trash uneaten."

She said about half of the students are obese "or at least moderately overweight" and some 75 percent of them have cell phones.

So why should they be receiving free breakfasts and lunches? She wonders.

"The school also provides day care centers for the un-wed teenage pregnant girls (some as young as thirteen) so they can attend class without the inconvenience of having to arrange for babysitters..." she said.

According to the teacher, students vandalized the new computers in the computer learning center, even though they were receiving a free education.

"I have had to intervene several times for young and substitute teachers whose classes consist of many illegal immigrant students here in the country less than three months who raised so much hell with the female teachers, call them *'Putas'* whores and throwing things... the teach-ers were in tears," she said.

She said the illegals in the district whose children are in the public schools believe they are entitled to be in this country but also demand certain rights, privileges and enti-tlements.

This has all led to "more crime, lower standards of edu-cation in our schools, overcrowding..." she said.

At the college level, illegal immigrant students qualify for PELL grants which they are not required to repay; for WAIT, a program that provides a credit card for the stu-dents for gas to go to and from school; and CARIBE, a pro-gram for illegals that pays for things like child care. U. S. citizens do not qualify for the CARIBE program.

I'm sure these students from the various foreign coun-tries are grateful for all of the men and women in this land who work long hours and pay heavy taxes to provide the gas credit cards and other financial programs for them.

The reason most illegals come to the United States is to find jobs. Companies hire them because they represent the "cheap labor" market.

But is the idea of "cheap labor" real or a myth? I believe it is a myth.

Here's the reason: illegals may only make seven or eight dollars an hour working for American companies, but they qualify for government housing and subsidized rent; food stamps; free (no deductible, no co-pay) health care; free breakfasts and lunches for their children along with bilingual teachers and books; financial relief from high energy bills; and for SSI (Supplemental Security Income).

"Cheap labor is cheap only to the employer," said U. S. Rep. Tom Tancredo of Colorado. "It costs the rest of us a fortune."

Some studies estimate that with all the government benefits available to illegals, a working father of five could be earning as much as $20 to $30 per hour.

And guess who's paying for this so-called "cheap labor"?

Parkland Hospital in Dallas, Texas, has the second busiest maternity ward in the nation where 16,000 new babies are born each year.

A patient survey conducted by the hospital revealed that seventy percent of the women giving birth at Parkland are illegal immigrants.

The Dallas Morning News reported that in one year the hospital spent $70.7 million delivering the babies. Medicate paid $34.5 million, Dallas County taxpayers $31.3 million and the federal government added another $9.5 million for other sources.

Federal law requires hospitals to provide medical care for illegals even though they cannot pay for the service.

According to the newspaper, the women who come to Parkland to have their babies also receive prenatal care including medications and can attend birthing and child care classes.

Car seats, baby bottles, diapers and baby formula also are available free of charge.

U. S. Rep. Michael Burgess of Dallas, who is an obstetrician and trained at Parkland, warned the people of Dallas

County "your property taxes are sky-high because you're paying for people who don't have medical insurance," the *Morning-News* reported.

The thousands of illegal women who come to Parkland realize they will receive free medical care.

It costs only two-hundred dollars to have a baby in Mexico.

"Many Parkland patients have acknowledged that they came to Dallas from Mexico or other parts of Latin America…" Sherry Jacobson said, writing for the *Morning-News*. "Some of the patients offer foreign passports and expired tourist visas as proof of those journeys. Others describe how they walked across the border and ended up in Dallas County."

Some of the illegal immigrant women are not pleased with Parkland because they don't have enough people in the maternity wards that speak Spanish and have threatened a lawsuit.

It seems to me that the inmates now are running the asylum!

The hospitals in Arizona and other border states taking care of illegals also are facing a financial crisis.

"Federal money set aside to reimburse hospitals… is going to run out at the end of the year," said Pamela Hughes of KTAR Radio in Phoenix.[3] "For the past four years the federal government has set aside $250 million a year to reimburse border hospitals…"

Dr. Michael Christopher, chief of staff of St. Joseph's Hospital in Phoenix said, "Indeed we spend millions upon millions each year, for the care of these patients."

"Christopher said in reality hospitals weren't getting much money from the feds to help cover those care costs," Hughes reported.

The doctor admitted the federals funds amounted to only pennies on the dollar but that it helped some.[4]

He added that hospitals all along the border have closed because they do not have funds necessary to take care of all the illegals, care forced on the hospitals by the federal government.

So be sure to pay your taxes, America, the illegal immigrants are depending on you!

A poll by a North Carolina-based immigration political action committee reveals that most Americans oppose taxpayer benefits for illegal immigrants.

Published by *American Pulse*, the survey found that four out of five are opposed to giving driver's licenses, business licenses, or housing assistance to illegals and three-fourths oppose financial support for education or job training.

William Gheen, president of Americans for Legal Immigration, says the mainstream media ignored the report and that the will of the majority of citizens in this land carries no weight with Congress.

"That means that the American population is being enslaved," he said. "Illegal immigration [is] rampant across the country and American taxpayers [are] getting stuck with the bills."

Ames Holbrook, who is a former deportation officer with Immigration and Customs Enforcement (ICE), has warned the American people that the federal government releases thousands of criminal immigrants every year because their own countries refuse to take them back after they have served their prison time, some for violent offenses.[5]

Hence, he says, they are free to roam the streets of the cities of this land.

Holbrook explained that as a deportation officer, he was required to ask the various countries to allow the United States to send the criminals back to their homeland.[6]

Those criminals include murderers, rapists, car thieves and child molesters.

Holbrook said the officials in those countries often

reply, "No, we won't take these criminals back, even if they are citizens of our country."

"Our own government...says, 'Okay, in that case, we'll release these felons here to our streets," Holbrook said.

Holbrook believes our officials should get tough with those countries. "We should say, 'Look, you're taking these guys whether you want to or not... we're going to make you take them back.'"

There are reports that Mexican soldiers are helping smuggle drugs into this country and engaged in an armed standoff with about thirty American law enforcement officers near Neely's Crossing about fifty miles east of El Paso. The crossing is well inside the United States.

The *Inland Valley Daily Bulletin* reported that Mexican military Humvees were towing vehicles loaded with what appeared to be thousands of pounds of marijuana across the border when they were challenged by the Americans.[7]

When the Mexican soldiers set up several machineguns, border patrol agents called for backup, *WorldNetDaily* reported.

"It's been so bred into everyone not to start an international incident with Mexico that it's been going on for years," Said Chief Deputy Mike Doyal of the Hudspeth County Sheriff's Department. "When you're up against mounted machineguns, what can you do? Who wants to pull the trigger first? Certainly not us."

FBI spokeswoman Andrea Simmons confirmed the afternoon encounter at Neely's Crossing. "People with Humvees, who appeared to be with the Mexican Army, were involved with the three vehicles in getting them back across (the river)," she said.

Doyal said the deputies captured one of the Humvees and found nearly fifteen hundred pounds of marijuana inside and the Mexican soldiers set fire to another that was stuck in the river. He said such incidents are common on the border.

"Last November (2006), his deputies were called on to back up agents from Fort Hancock border patrol station in Texas after confronting… fully armed men in Mexican military uniforms," *WorldNetDaily* reported. "Armed with machineguns, the men were trying to bring more than three tons of marijuana across the border in military vehicles."

However, Michael Chertoff, then-director of Homeland Security, was skeptical of the reports of Mexican soldiers coming across the Rio Grande into the United States. And Mexican officials denied that any of their soldiers had ever crossed into this country.

A new documentary film titled *Border* presents a shocking but true picture of the problems created by illegal immigration along this country's border with Mexico. The film was directed by Chris Burgard, an actor who has appeared in several television programs including "JAG" and "Growing Pains."

"'Border' takes you on a journey from state to state and exposes a failed system and a failed policy," the film's website says. "Burgard's courageous journey includes powerful footage in the border crossing regions of America where dead bodies, armed Mexican military incursions… and drug traffickers are commonplace."

In an interview with *FrontPage* magazine, Burgard said he began the film project after hearing stories of some of the terrible things that were happening along the border. "I first started hearing about the rapes on the border and how the *drogandos* would put guns to people's heads and say, 'You are going to carry this backpack of drugs into America or we are going to kill you and leave you in the desert…'"

"The film is shot largely along the border and contains images of dead and decaying illegal immigrants, many that local ranchers say were murdered by human traffickers…," *NewxMax.com* reported.

Burgard said the situation is so critical that some ranchers are living behind razor-wire fences around their homes and others are on the verge of bankruptcy because of vandalism by the illegal immigrants and drug smugglers coming across the river.

"Dead bodies are so numerous that border counties are going broke just paying for all the coroner inquests," Burgard said.

Sheriff Larry Dever of Cochise County, Arizona, says that ten percent of the illegal immigrants they catch coming across the border already have felony convictions in the United States.[8]

According to the sheriff, some of those are looking for jobs but others are hauling tons of drugs into this country.

Officials estimate that each day a thousand illegals enter the United States through Cochise County which has an 83-mile border with Mexico.

"Frankly the biggest burden has been to the community," Sheriff Dever said. "The disruption of normal social life. People ... stay home and watch their place so it's not invaded, burglarized..."

It is quite apparent to anyone who has even a basic understanding of the chaos on the Rio Grande that those in government from the courthouse to the White House are doing virtually nothing to solve this critical problem of the invasion from the south. And I sometimes wonder why.

Although all of the illegal immigrants are in this country in violation of federal law, some Hispanic leaders in this land are speaking out against the American citizens and for the illegals. Here are some of their statements as reported by the California Coalition for Immigration Reform:

Augustin Cebada, of the Brown Berets, said, "Go back to Boston! Go back to Plymouth Rock, Pilgrims! Get out! We are the future... we have beaten you... we are going to take over."

Richard Alatorre, the Los Angeles City Council,

"They're afraid we're going to take over the governmental institutions. They're right. We will take them over."

Excelsior, the national newspaper of Mexico wrote: "The American Southwest seems to be slowly returning to the jurisdiction of Mexico without firing a single shot."

Professor Jose Angel Gutierrez of the University of Texas said: "We have an aging white America... They are dying. The explosion is in our population... I love it."

Mario Obledo, California Coalition of Hispanic Organizations, said: "California is going to be a Hispanic state. Anyone who doesn't like it should leave."

On May 5, 2010, a day known as *Cinco de Mayo*, five young men in the Live Oak High School in Morgan Hill, California, wore American flag clothing to school. The principal ordered them to turn their shirts inside-out to conceal the flag. When they refused, he suspended them and sent them home.

The superintendent of schools in the district later apologized to the five boys and their parents for the incident and ordered a review of the principal's action. However, the next day some two-hundred Mexican-American students, carrying Mexican flags, walked out of class.

Also, the Klein High School in Houston, Texas, also celebrated *Cinco de Mayo* in 2010 and flew the Mexican flag. A white student removed the flag and was suspended for three days.

It is my understanding that *Cinco de Mayo* celebrates the Mexican victory over Emperor Maxmilian and his French soldiers. That, no doubt is very important to the Hispanics living in this land.

But flying the Mexican flag in our schools? This is the United States, not Mexico.

During a rally for illegal immigration in Maywood, California, supporters took down an American flag, stamped on it and replaced it with the flag of Mexico. When police officers removed the Mexican flag, the people threw bottles and rocks at them.

Students at the Velasco Elementary School in Houston, Texas, celebrated a "diversity" assembly by saying the pledge of allegiance to the Mexican flag. Radio Station KTRH posted a clip showing the students cheering and waving the Mexican flags. Parents voiced their outrage over the event.

A cameraman for a Houston television station placed a Mexican flag on his camera while covering a rally in support of illegal aliens. His flag drew angry shouts from counter-demonstrators.

According to the *Houston Chronicle,* hundreds marched to Houston's Mason Park in support of the illegals. The newspaper reported those marching blew whistles, banged on drums and chanted as they carried both American and Mexican flags.

Meanwhile, Mexican President Felipe Calderon accused the United States of persecution of illegal immigrants from Mexico.[9]

"I want to express again an energetic protest at the unilateral measures taken by the U. S. Congress and government which exacerbate the persecution and abusive treatment of undocumented Mexican works," Calderon said.

Mexico reportedly is disappointed that the congress did not pass the amnesty bill, even though President Bush strongly supported the measure and President Obama strongly supports the measure today. The country also has expressed its opposition to any kind of border fence between the two countries to keep the immigrants out of the United States.[10]

Of course, President Calderon has certain ulterior motives. He knows that if poor workers from Mexico enter the United States it will mean less welfare payments by his government. Also, those workers send millions of dollars back to Mexico to family members each month and that money is important to the Mexican economy.

But President Calderon has a double standard regarding immigration – one for the Unites States, another for his own country.

Mexico has one of the toughest sets of immigration laws anywhere in the Western Hemisphere and, I suspect, most Americans would like to adopt their policies rather than consider amnesty for the millions of illegals in this land.

In Mexico it is a felony to be an illegal alien.[11]

Here are some of the components of Mexican immigration law: all aliens must be in the country legally and have means for self-support and not be a burden on the taxpayers of Mexico. Any of the aliens found to have criminal records or to have entered the country illegally are deported immediately.[12]

The Mexican government prohibits aliens from involvement in politics and those who violate that regulation are imprisoned or deported.

Those regulations have far-reaching implications for all foreigners living in Mexico.

For instance, foreigners are not provided welfare, food stamps, health care or any other help by the government. They are not allowed to hold demonstrations against Mexico and cannot fly the flags of any other countries. Those who do so are sent home.

And they have zero-toleration for any foreigners who enter Mexico illegally.

The naturalized citizens and the illegal immigrants appear to becoming more and more antagonistic toward the United States.

When the United States soccer team met the Mexican team in the Los Angeles Coliseum, the pro-Mexican crowd booed the National Anthem and held American flags upside down.

"Supporters of the U. S. team were insulted, pelted with projectiles, punched and spat upon," the California Coalition said. "Beer and trash were thrown at the U. S. players before and after the match."

Now let's summarize the illegal immigration situation.

About one-third of the... illegals in this land are on some form of welfare and that costs the American taxpayers about twenty-billion dollars a year, according to a report by the Center for Immigration Studies and based on Census Bureau statistics.

Mark Krikorian, the executive director of the center, said that the negative aspects of continued immigration includes "an increasing burden on taxpayers to subsidize a 19th century workforce imported into a 21st century society..."

"Immigrants who have legal status, but little education, generally have low incomes and make heavy use of welfare programs," the Immigration Studies report said. "If we decide to legalize illegal immigrants, we should at least understand that it will not result in dramatically lower welfare use or poverty."

The report also pointed out that those who support amnesty should not try to convince the American people that amnesty will lower taxes and result in higher incomes for the illegal immigrants.

"Legalized aliens will still be overwhelmingly uneducated and this fact has enormous implications for their income, welfare use, health insurance coverage, and the effect on American taxpayers," the report said.

Krikorian emphasizes there "is no excuse for intolerant attitudes toward legal immigrants – we admitted them according to the rules established by our elected representatives, and we must, and will, continue to embrace them as Americans in training."

"Even illegal immigrants must be treated humanely as they are detained and returned to their homes," he said. "But future legal immigration is a different question – mass immigration is simply not compatible with the goals of modern society and should be minimized to the extent possible."

Now I want to present the BILL KEITH-NO NON-SENSE-SOLVE THE DILEMMA-PLAN for immigration reform.

There are two important underlying principles we must recognize: First, the immigrants from the south need jobs, and, second American companies need workers.

Since a vast majority of the American people are opposed to amnesty for the twelve million illegal immigrants in this land, how do we solve the problem?

The illegals that I know personally are good and decent people with a strong work ethic. But I am aware there are others such as the freeloaders, drug dealers, gang members and the those who come across the border every day just to have their babies in our hospitals.

Here's my plan.

First, all workers currently in the United States, who are employed, will be issued a TR (temporary resident) card that will be valid and give them legal work status as long as they continue to work and pay taxes. It would be imperative that these cards be tamper-proof for some illegal engravers have become experts at preparing false identification papers of all kinds.

Should the workers with the TR cards resign from their jobs or be terminated, they would have a ninety-day grace period to find another job. However, if they are unsuccessful in new job location, their card would no longer be valid and deportation workers with ICE would return them to their homeland.

Second, all employers would be required to provide health insurance for these workers and their families. And, the workers would be required to pay their fair share of that insurance. That would greatly relieve the exorbitant costs of taxpayer-funded medical assistance to the millions of illegals in this land.

I have personal knowledge of some illegal workers who choose not to participate in the company insurance plan for they know that free government medical assistance is available to them and their families.

Third, the government and local police authorities would crack down on all day-labor programs where employers pay the illegals in cash at the end of the day and the employees, in turn, pay no taxes. Many employers like this arrangement since they are not required to pay for health insurance or any other benefits.

That should not be tolerated. The employers, who receive the benefit from the workers, should pay for their medical expenses and not push the cost off onto other taxpayers.

Fourth, all other illegal immigrants would be returned to their native lands.

Fifth, the president of the United States should immediately post National Guard units all along the border with Mexico to turn back the estimated four thousand illegals who are coming into this country every day.

I know that the *posse comitatus* law limits the use of the military for civilian law enforcement. But we have been fighting a war against terror since 9/11 and, in my opinion, that would justify using troops to protect our border for we are facing a disastrous situation that may have grave consequences for this nation.

Obviously ICE, working with the employers, would be responsible for monitoring each of the TR workers. But that would be far superior to the ineffective manner those workers are being handled today.

My plan may be much too simple for congress since they are masters at creating problems and then campaigning against them. I'm sure they would want something far more complicated.

But this plan will work.

I am sending this proposal to the White House and to my two United States senators and my congressman.

Perhaps someone will listen and try to make some sense out of a lot of nonsense.

No one can blame the illegals for wanting a better life in America. However, the good, decent, hard-working Mexicans – and those from other Central and South American countries – are not the only ones crossing the river into what they believe is the "Promised Land." There are others such as drug pushers, criminals, gang members and, God forbid, terrorists.

Illegal immigration is a problem, but not the biggest problem. The most serious problem we face is our dysfunctional government's inability to deal with illegal immigration. Those in government, democrats and republicans, don't have a clue how to remedy the problem.

Perhaps they should try my plan.

Chapter Eleven
American Tragedy

There is a culture of death hanging like a wreath around the American conscience, the result of 50-million unborn babies killed since 1973 when the U. S. Supreme Court approved abortion on demand.

We have lost the value of human life as millions of would-be mothers have voluntarily given up their unborn before they drew their first breath. This has contributed to a syndrome of death that has spread through movies, television and violent video games. Death has become so common that people's consciences have been anesthetized to the value of human life and death is just a bullet or a stab-wound away.

Pope John Paul II warned us about the "new forms of attack on the dignity of the human being." He said that "the end result is tragic: conscience itself, darkened as it were by such widespread conditioning is finding it increasingly difficult to distinguish between good and evil in what concerns the basic value of human life." This leads to a culture of death, he said.

Of all the absolutes accepted as dogma by the secular progressives, abortion is the one I find most alarming. It is difficult for me to understand how most in the news media, academia, Congress and the United States Supreme Court support the infanticide of helpless unborn infants resting in their mothers' wombs waiting to see the light of day. Since the high court ruled in Roe v. Wade in 1973, a mother's womb has become the most dangerous place on earth, more dangerous than the mine fields of Iraq or Af-ghanistan, the Muslim suicide bombers, Hurricane Katrina, the *tsunami* that devastated Sri Lanka, the earthquake in Haiti or the gangs of New York City and Los Angeles.

Through the years I have often asked: "What drives this madness?"

Does anyone know for sure?

We hear all the mumbo-jumbo about "a woman's right to choose," "coat-hanger abortions in back alleys," and "rape and incest."

However, surveys reveal that 97 percent of all abortions are for birth control or convenience.

Years ago I got into an argument with an ACLU attorney in the news room of *The Shreveport Journal,* where I was the city editor, on the subject of abortion. I argued that abortion was the senseless killing of the unborn child that could not defend himself.

The attorney argued that when his girlfriends became pregnant, they couldn't party and have fun at night. So they had abortions. In their relentless pursuit of ecstasy, a cheap orgasm became more important than the life and death of the unborn.

But I believe there are more sinister answers to the question of why people support this ghastly procedure.

I'm convinced that abortion often is based on selfishness.

For instance, some women just don't want to be bothered by birth-control measures or by a child.

The answer? Abortion.

Others believe the earth is becoming over-populated and there will not be enough resources such as food to care for all of them. They fear that too many people will be a burden on government spending on health care and welfare benefits. Hence, they want to control the population.

There are 6.6-billion people in the world. When we do the math, we see there is enough land in the three countries – the United States, Canada and Australia – for every person in the world to have an acre of land.

Hence, the overpopulation argument is somewhat ludicrous. Also, the farmers in the Western World and Australia could raise enough food to feel those billions and wipe out hunger forever.

Why are we not doing it?

There may be another reason why so many people have no reservations about abortions. The Bible teaches us that some have violated their innate conscience so often that they have become deadened like being "seared with a hot iron" and are without feeling.

What a tragedy!

I've long believed that the loss of conscience and the death of the meaning of life results in serious problems, particularly for the women who have been the victims of pro-choice propaganda.

Now comes a report from the Royal College of Psychiatrists that abortion is a risk to mental health.

"Several studies, including research published in the *Journal of Child Psychology and Psychiatry*, concluded that abortion in young women might be associated with risks of mental health problems," wrote Sarah-Kate Templeton, health editor, in the *Sunday Times* of London.[1]

The tragic death of a talented artist in Cornwall in 2008 has drawn attention to the abortion/mental illness dilemma. The artist hanged herself because she was overcome with grief after she chose to abort her twins before they were born, the *Sunday Times* reported.

Emma Beck, thirty, left a suicide note that said: "Living is hell for me. I should never have had an abortion. I see now I would have been a good mum. I want to be with my babies; they need me..."[2]

Several members of England's parliament are seeking a "cooling off" period where women seeking abortions would be informed of the "possible consequences of abortion, including the impact on their mental health" before they have the procedure."[3]

"For doctors to process a woman's request for an abortion without providing the support, information and help women need at this time of crisis I regard almost as a form of abuse," said Nadine Dorries, a Conservative member of Parliament who supports the "cooling off" period.[4]

Christian leaders across this land have issued a warning: Unless we who believe in the sanctity of life wake up and take a stand for life, we will lose everything we have worked for during these past three and a half decades to save the lives of the unborn.

"Beyond losing the chance to restore protection to unborn children in our lifetime, we face losing every single pro-life law we have passed in the last 34 years," writes Janet Folger, a Christian activist and author. "Parental notice, parental consent, the woman's right to know law, waiting periods, fetal homicide, abortion funding restrictions, partial-birth abortion bans – all gone – wiped off the books in every single state…" she said.

Folger pointed out that the secular progressives in congress are pushing the radical "Freedom of Choice Act" and, if it is passed, all that will be required for it to become law is President Obama's signature.

"Not only did they pass 'hate crimes' legislation which, without a presidential veto will criminalize Christianity, now nearly a fourth of the U. S. House of Representatives and a fifth of the U. S. Senate are cosponsors of the most radical abortion bill this nation has ever seen," she said.[5]

Folger points out that if the act becomes law the people of this land can "say goodbye to every pro-life advance we have made… and say hello to partial-birth abortion performed on your 12-year-old daughter without your knowledge or consent, paid for with your tax dollars in every single state."

Many conservatives in Washington are reluctant to speak out against the "Freedom of Choice Act" for fear of being crucified by the news media. And even many of the preachers in our pulpits are silent on this and other critical issues.

But there were five notable exceptions: Pope John Paul, Mother Teresa, pastors Jerry Falwell and D. James Kennedy and Congressman Henry Hyde of Illinois. But now all of them are deceased.

Where are today's voices crying out on behalf of the unborn? Who will take up the mantle of these great men and the courageous Mother Teresa?

One such voice is Dr. Alveda King, niece of Rev. Martin Luther King, Jr., the civil rights leader who was assassinated, who believes the death of the unborn is a burning issue today as was civil rights during her uncle's generation.

"In s speech at a church in downtown Memphis, King noted how her uncle was killed... in 1968," said Steven Ertelt, editor of LifeNews.com. "But a short five years later the Supreme Court opened the floodgates to unlimited abortions in the Roe v. Wade decision – one that King says has decimated the nation and the black community in particular."

She believes that abortion is the new frontier of this land's current civil rights movement. Speaking of her uncle's stand on abortion, he encouraged everyone to "admonish black women not to abort their babies" but added "we don't show them how to care for them, don't give them options for life."

Alveda King says she is ashamed of all the black leaders today who endorse abortion.

"If we're not speaking for the voiceless, not being strong for those who are weak, we're not living Martin's dream," she said.

There are several wonderful organizations that speak

out on behalf of the unborn such as Focus on the Family, American Family Association, Concerned Women of America, the Catholic League and some others.

For nearly half a century we have seen the innocent unborn destroyed. And many of us have given mental assent to law that allows this American tragedy. But what is it to which we have given assent? Is it not a holocaust of giant proportions and even a civil rights disgrace all in one? Those things once universally accepted as criminal and rejected by decent people now has, as Pope John Paul II said, "become socially acceptable."

Feminists, secular progressive politicians and the news media have conditioned the people of this land to accept the lie that abortion is socially acceptable.

If wombs had windows, people would see the hurt and pain inflicted on the helpless, defenseless unborn and we would recoil in distaste. The men of this land – particularly the hard hats – would rise up and demand that no such procedures for convenience ever be performed again in this land. They would run the doctors who perform these ghastly procedures out of town and tell them never to come back.

But does anyone really care?

We are far worse than Adolph Hitler's Germany where six million Jews died in the concentration camps. Hitler and his Gestapo legions killed them in the gas chambers and death ovens. In America we kill them in their mothers' wombs. The Germans' death chambers were in Auschwitz, Ravensbruck and Dacau. Ours are in the fashionable doctors' offices in the New York City, Cleveland, Atlanta and all across the land.

We have birthed in this land a new generation of doctors who destroy the unborn without mercy and mothers who have been brainwashed into believing that evil is good and aborting their baby is the right thing to do.

Ironically, we have a former president, members of the

U. S. Supreme Court and Congress who sit in church every Sunday who are totally obsessed with getting rid of the unborn.

Why do the men in our pulpits not cry out against this holocaust? Because it just isn't politically correct and they are concerned more about what people think than what God thinks.

Reason has not worked. Access to the political process has not worked. Demonstrations have not worked. Neither secular progressive democrats nor marshmallow republicans in Washington have taken any serious stand on behalf of the unborn. Of course, President George W. Bush was a shining exception to that. But during the last two years of his administration, he ran the White House. The progressives ran Congress.

Our laws break all the rules of reason. They protect the right to kill, not the right to live.

The socio/political banner that drives this madness is called freedom of choice. Actually it is the right to kill and maim and inflict pain and destroy the unborn who has no voice and no choice. And the courts protect those who kill and jail those who try to save lives.

Planned Parenthood, the biggest abortion provider in the land, receives an annual federal subsidy of three-hundred-million dollars. There is some evidence the organization performs late-term abortions in violation of the laws of most of the states in the land and do not determine the viability (if the baby can live outside the womb) of the babies before the abortions.

According to Troy Newman of Operation Rescue, abortion clinics are "the most unregulated, unaccountable places in the United States."

Newman says it is inappropriate for Planned Parenthood to receive millions of dollars of taxpayers' money when it is unclear whether they are breaking the laws of various states.

Rev. Joseph Parker, a black pastor of the Campbell Chapel AME Church in Pulaski, Tennessee – birthplace of the Ku Klux Klan – has also spoken out against Planned Parenthood.

He told his congregation that Planned Parenthood and other abortion providers are much more dangerous than the Klan because the abortion provider takes the lives of more blacks than violence, accidents and all diseases combined.[6]

Reverend Parker said that most would agree that the Klan is the number-one enemy of blacks. But he notes that all the violence the Klan has done to blacks over the years "pales in comparison" to the violence against unborn black babies by Planned Parenthood.

"Planned Parenthood kills more black people in three days than the Klan has killed in its entire history of existence," he said. "Yet many in the African-American community [who] don't even see Planned Parenthood as an enemy would see them as a friend. So one of the things we're using as a platform is explaining that 'you may think the Klan in your enemy – let me show you the real enemy.'"

As I mentioned earlier, Planned Parenthood used Mother's Day as an event for fundraising.

In our troubled culture, unborn children and the disabled are being thrown away as though they have no value, says Will Hall writing for Baptist Press.

Our culture has a "declining esteem for human life, especially the weakest among us," Hall said. "American freedom used to be defined in terms of 'life, liberty and the pursuit of happiness,' but now more often than not it is talked about in terms of a right to die and a right to an abortion...

"Also, although recent data indicates America is experiencing a declining abortion rate, the killing of the unborn continues as a plague in the U. S. – on average three-thousand-five-hundred children in the womb are killed each day."

Hall says there is a "troubling and increasing trend in America to devalue or at least 'selectively' value the unborn."

He adds that some believe abortion to be in the best interest of the baby.

Wonder what the baby would think about that?

Hall also pointed out an alarming trend: That at least half of the abortions are the second, third and fourth for the women.

Hall says that every human being has the spiritual capacity to fellowship with God. "Consequently, seeing any human, the crown of His creation, as something less than a spiritual being, denies the measure of every man's true worth..." Hall said. "The sum of man is not the total of his physical experiences; rather, his value was fixed by God from the beginning as something He treasures, so much so that He sent His Son to die for all men in all our various defects, in all our spiritual corruptions."

Now comes word that the American College of Obstetricians and Gynecologists (ACOG) has issued new guidelines that require physicians who choose not to perform abortions to refer their patients to a doctor who will do the procedure.

The ACOG, which certifies doctors, apparently has the power to refuse or revoke certification for doctors who do not follow the new ethics rule.[7]

Wendy Chavkin, an Ob/Gyn from Columbia University was pleased with the new ethics statement.

"It says that if a physician has a personal belief that deviates from evidence-based standards of care they have to tell the patient that, and that they do have a duty to refer patients in a timely fashion if they do not feel comfortable providing a given service," Chavkin said.[8]

However, some doctors disagree with the new ruling.

"I'm not going to refer someone to a hit man to put to death someone that's inconvenient in their life," said Joseph DeCook, a retired Ob/Gyn from Holland, Michigan. "I wouldn't do that... I'm not going to refer a pregnant woman to a physician who will purposefully terminate her pregnancy..."

Do you see what is happening? It's conceivable that Christian doctors who believe in the sanctity of life and refuse to perform abortions could conceivably lose their licenses for protecting the life of the unborn.

The Bush administration's Health and Human Services challenged the new policy issued by the ethics committee of ACOG.

"We had great concerns that technically competent, skilled, highly trained physicians could be denied board certification solely on the basis of refusing to refer for abortions, something that might be against their moral compass or ethical stand," said Don Wright, a deputy assistant secretary.

There may be some light at the end of the tunnel.

Current surveys reveal that for the first time since Roe v. Wade, in 2010 some fifty-one percent of the American people now oppose abortion on demand.

Also, Focus on the Family reports that most young Americans hold conservative views on the subjects of abortion and same-sex marriage. The report is based on a poll by *The New York Times,* CBS and MTV.[9]

Sixty-two percent of the seventeen-to-twenty-nine-year-olds surveyed, said abortion should be outlawed or restricted.[10]

Danielle Huntley, a law student at Boston College and president of the Students for Life of America, said she is pleased to learn that students are not following the dogma of the secular progressives.

"It illustrates that my generation realizes that they are survivors of Roe," she said. "Each of us born after 1973 could have been legally aborted by our parents."

Tom Robins of the College Republican National Committee said, "Our generation has seen the effects of that... They understand that abortion is not a healthy choice for America."

Also, fifty-four percent of the students in the poll revealed their opposition to same-sex marriages.

Hooray for the young people! Apparently they have much better judgment than most of the secular progressives in the land.

Don Feder, who for two decades wrote political and other columns for the *Boston Herald*, once said: "The outcome of our national debate about abortion, homosexuality and drugs will determine what sort of people we become. Immigration, English, patriotism, multiculturalism and national security go to the heart of whether we will remain a people, in a coherent sense, at all."

Whether we like it and whether we even want to think about the senseless deaths of the unborn, we must be the voices of the voiceless, the hope of the hopeless. We are the ones who must cry for those whose cries will never be heard.

Chapter Twelve
World War III?

On Dec. 21, 1988, Pan American Flight 103, a Boeing Jumbo Jet known as the "Clipper Maid of the Seas," departed Heathrow International Airport in London bound for John F. Kennedy International Airport in New York City. Only 38 minutes after departure and flying at 31,000 feet, the plane exploded in mid-air over Lockerbie, Scotland, killing all 259 passengers and crew and 11 citizens of Lockerbie.

"Within three seconds of the bomb detonating, the cockpit, fuselage and No. 3 engine were falling separately out of the sky..." wrote author Hugh Miles for the *London Review of Books*.[1] "With the cockpit gone, the fuselage depressurized instantly and the passengers in the rear section of the aircraft found themselves staring out into the Scottish night air."

Miles said that rescue teams found dead passengers "clutching crucifixes, or holding hands, still strapped into their seats."

Investigators from the Dumfries and Galloway Constabulary and the Federal Bureau of Investigation in the United States examined 10,000 pieces of debris scattered over an 81-mile area and found fragments of a circuit board that detonated a bomb made of one pound of plastic explosives. They determined that the bomb had been hidden inside a Toshiba Radio and smuggled onto the plane in a Samsonite suitcase.

After months of meticulous police work, they traced the

bomb to Abdelbaset Ali Mohmed Al Megrahi, an Islamic terrorist and former Libyan army intelligence officer.[2] On Jan. 31, 2001, three Scottish judges convicted Al Megrahi of the crime and sentenced him to life in prison. In 2009, after serving only eight years, the judges released him for health reasons and he returned to Libya to a hero's welcome.

There had been several previous terrorist attacks against Americans, but the wanton killing of the passengers of Pan Am Flight 103 signaled a new era in the Islamic militants' war on the people of the United States. Let's take a look at the terrorists' trail of blood:

- On November 4, 1979, Islamic extremists – members of the Revolutionary Guard – attacked the American embassy in Tehran, Iran, and captured fifty-two American personnel and held them captive for 444 days. Ayatollah Khomeini, the supreme commander of the land after the fall of the Shah of Iran, refused to release the Americans until shortly after Ronald Reagan became president in January of 1981.

- In April of 1983, terrorists attacked the American Embassy in Beirut, Lebanon, and killed 17 Americans.

- In October of 1983, terrorists attacked the United States Marine barracks in Beirut and killed 241 American Marines. Early in the morning, a delivery-type truck approached the barracks, drove through a barbed-wire fence and rammed into the barracks. The terrorist driver detonated t12,000 pounds of TNT that completely destroyed the four-story building where the Marines were sleeping. "It was the bloodiest day in the Corps' history since World War II, when the Marines sought to secure Iwo Jima," said Sgt. Melvin Lopez Jr., writing for the *Henderson Hall News*, a military newspaper.

- On March 16, 1984, William Buckley, CIA chief in Beirut, Lebanon, was kidnapped by the Iranian-backed Islamic Jihad. They tortured then executed him.
- In August of 1985, a car bomb explosion at an American military installation in Frankfurt, Germany, killed two soldiers and injured 20 others.
- In October of 1985, Palestinian terrorists' attacked the Achille Lauro cruise ship. They killed a tourist in a wheelchair by the name of Leon Klinghoffer, an American of Jewish heritage. They shot him and threw his body overboard.
- On April 5, 1986, terrorists bombed the LaBelle discotheque in West Berlin. The bomb killed one American and wounded forty-four others. Libyan terrorists took credit for the bombing.
- In April of 1986, terrorists bombed a Trans World Airways flight preparing to land in Athens, Greece. Four passengers were killed when a vacuum of wind pulled their bodies through a gaping hole in the fuselage.
- On Feb. 24, 1993, terrorists detonated a massive bomb at the World Trade Center in New York City killing six and wounding 1,042.
- On August 7, 1998, Al-Qaeda terrorists simultaneously bombed the United States embassies in Nairobi, Kenya and Dar es Salaam, Tanzania. Hundreds were killed.
- On Oct. 12, 2000, Islamic suicide bombers attacked the USS Cole, a navy guided missile destroyer docked in the Yemeni Port of Aden to refuel, and killed 17 American sailors. The terrorists rammed the port side of the Cole and the explosion blew a large hole in the side of the ship.

According to the Center for Defense Information in

Washington, D.C., Islamic terrorists increased their *jihad* or holy war against the United States in the 1980s. Their crusade reached a horrendous climax on 9/11 when Al-Qaeda terrorists flew two Boeing 767 jets into the twin towers of the World Trade Center in New York City killing two-thousand-seven-hundred and fifty-two people.

After Libyan terrorists bombed the LaBelle Discoteque in Paris on April 16, 1986, President Ronald Reagan retaliated by sending American Air Force jets to bomb key military installations in the capital city of Tripoli, Libya.

Hence, President Mohamar Kadhafi of Libya decided not to mess with Reagan and the United States and terrorist activities in his country virtually ceased.

Certain other presidents probably would have said something like this: "What we need to do is sit down and talk to Kadhafi, we know he is a good man and will listen to reason."

Reagan had a different idea and it worked.

It took us a long time to catch on to the reality of the Muslim terrorists' war against our people and our land. President George W. Bush was the first leader who had the guts to go after our enemies overseas rather than waiting to fight them in our homeland. He sent troops to search out and destroy their desert hideouts and training camps in Afghanistan and Iraq.

On October 7, 2001, British and American military units attacked the Taliban and Osama bin Laden's Al-Qaeda strongholds in Afghanistan. Five weeks later the coalition forces marched into the capital city of Kabul as the Taliban and bin Laden ran for the hills.

Then on March 20, 2003, American, British, Polish, Australian and Danish troops invaded Iraq, where thousands of Al-Qaeda operatives were hiding and within weeks toppled Saddam and took control of the country and opened the doors for freedom and democracy.

Today the heroic men and women of the Unites States military are paying a heavy price for the do-nothing past presidents and members of congress who should have challenged the Islamic terrorists years ago, but did virtually nothing.

On other Islamic terrorist matters, I remember some pictures of Muslims marching through the streets of London in August of 2007 during a "Religion of Peace Demonstration." They carried posters that said:

"Butcher Those Who Mock Islam"

"Europe, You Will Pay – Demolition Is on the Way"

"Islam Will Dominate the World"

"Freedom Go to Hell"

"Be Prepared for the Real Holocaust"

That's some convoluted "peace" demonstration. So much for the "peace-loving" Muslims!

However, it troubles me that these pictures would never be shown on American television as they might encourage the American people to support the war on terrorism, a war the media almost unanimously oppose.

These media elite, the ACLU, the secular progressives in Congress, some federal judges and the self-love Hollywood crowd also oppose the Patriot Act that gives the government freedom to place suspected terrorists under surveillance and listen to their telephone conversations, monitor their email transmissions and a host of other tools being used in the war on drugs and organized crime. Those groups are doing everything possible to take away our safeguards under the banner of civil liberties.

Islamic radicals say they will conquer Europe and England will be the first to fall. Yet the English allow thousands of Muslims into their land each year.

Apparently several Muslim doctors living in England were practicing more than medicine when they made plans to bomb the Tiger Tiger Nightclub in London.

ABC News reported the bomb plot failed because of a

malfunction in the firing mechanism that was placed inside a Mercedes E 300 parked near the front door of the club.[3]

"Had the fuel-air bombs successfully ignited into a superhot fireball filled with roofing nails, casualties were almost a certainty among the 500 or so patrons who partied late at the 1,700-person occupancy nightclub that perhaps best symbolizes London's vital nightlife scene," said Richard Esposito and Jim Sciutto reporting for ABC News.

However, Esposito and Sciutto said that an ambulance crew, in the vicinity on an unrelated matter, saw "a plume of cold propane" coming out of one of the windows of the car.

"When a bomb technician in a 90-pound Kevlar suit walked down to the vehicle to examine it, he also found a firing system rigged inside the car and another inside its trunk along with four jugs of gasoline," Esposito and Sciutto said. "The technician successfully disarmed the devices."

The reporters also said that a second Mercedes also rigged with a bomb device was parked a few hundred yards away from the entrance to the nightclub. They speculated the terrorists planned to blow it up to kill the patrons who escaped the first explosion.

"Within 14 hours after the plot failed, the same two men believed to have planted the bombs in London attempted what appears to have been a suicide incendiary attack on the doors to a terminal at Scotland's Glasgow Airport," Esposito and Sciutto said. "That attack failed too. The vehicle failed to reach the doors, and its contents failed to ignite even after one of the occupants tried to douse the car in gasoline, setting himself on fire in the process."

Immediately after the attempted bombing, investigators from Scotland Yard arrested eight people, five of them doctors, *The Independent* newspaper of London reported.

"A suspected secret cell of foreign militants, believed to be linked to al-Qa'ida and using British hospitals as cover, are being questioned..." Kim Sengupta, Ian Herbert and Cahal Milmo reported for the paper.

According to the newspaper, most previous incidents of Islamic terrorism involved Muslims who were citizens of England.

"The alleged arrival of teams from abroad to carry out attacks, their identities unknown to the domestic law agencies, adds another dimension to the terrorist threat being faced in the United Kingdom," the reporters said.

Police raided several homes of suspected terrorists and arrested foreign nationals from Jordan, Saudi Arabia and Iraq.

"Among those arrested was Mohammed Jamil Abdelqader Asha, a 26-year-old neurologist who was born in Saudi Arabia but is of Palestinian origin and was traveling on a Jordanian passport," the reporters said. "He and his 27-year-old wife, a medical assistant, were arrested... in connection with the attempted bombings in London."

They also arrested Bilal Talal Abdul Samad Abdulla, an Iraqi citizen from Bagdad who came to England in 2006, *The Independent* reported.

"He is said to be one of the two men... in the Glasgow airport attack, and is suffering from third-degree burns," the reporters said.

Five of the men who planned the abortive attack on the Tiger Tiger Club were convicted of acts of terrorism and sentenced to forty years in jail.

What drives this madness?

Fox News reported that a well-known Islamic *imam* and a lawyer in London support killing and rape of non-Muslims.

"A question-and-answer session with Imam Abdul Makin in an East London mosque asks why Allah would tell Muslims to kill and rape innocent non-Muslims..." Fox said.

He answered: "Because non-Muslims are never innocent, they are guilty of denying Allah and his prophet," the *imam* replied.[4]

The lawyer, Anjem Choudray, supports the *iman's* position. "You are innocent if you are a Muslim," he said in an interview with the British Broadcasting Company. "Then you are innocent in the eyes of God. If you are not a Muslim, then you are guilty of not believing in God."

"As a Muslim, I must support my Muslim brothers and sisters," Choudray said. "I must have hatred to everything that is not Muslim."

I am told there are limited nuclear devices for sale on the world's black market that are small enough to fit into a suitcase. I wonder what it will take for the American people to realize the threat from Islam to this land. Perhaps a mushroom cloud over one or more of our major cities?

The Daily Telegraph of London reports that security agents in Colombia have recovered sixty pounds of depleted uranium from leftist rebels apparently planning to make a so-called "dirty" bomb.[5]

Although depleted uranium cannot be used to make a nuclear bomb, added to various explosives it can spread dangerous radiation once exploded, the newspaper said.

Now let's put two and two together and take a look at a very dangerous scenario. We have thousands of illegals coming into this country every day from Mexico and other countries as far away as China. What if sixty pounds of depleted uranium was smuggled into this land by a terrorist, added to a mix of diesel fuel and ammonia nitrate – the ingredients in the bomb that destroyed the Alfred P. Murrah Building in Oklahoma City on April 19, 1995 – and detonated it in one of our major cities? It would not have the effect of an atomic bomb but the radiation would cause panic, force people to evacuate their homes and it would be months or years before the area would be free of radiation. Also, high winds could carry that radiation to other areas of the country.

We Americans are living in a comfort zone seeking hedonistic pleasures and possessions and don't want to be bothered with the thought of any kind of calamity or destruction such as 9/11. But while we are lulled to sleep by false security, our enemies work day and night planning our destruction.

"The public is getting complacent," Ray Kelly, police commissioner in New York City, told Fox News.[6]

Fox reported that security experts have issued a warning against serious terrorist attacks as a number of cases of so-called "homegrown terrorists" or American jihadists enter the federal courts.

"Several terror-related cases now in the courts highlight this need for continued vigilance..." Fox reported.

One case involved the so-called "Liberty City Seven" who planned to destroy the Sears Tower in Chicago. FBI agents uncovered the plot through secret surveillance tapes and discovered the terrorists had close ties to Al Qaeda.[7] In the State of Washington, a Pakistani-American by the name of Naveed Haq was on trial for killing one and wounding five others at Seattle's Jewish Federation Building.

Another was the trial of Houssein Zorkot, a native of Lebanon and a medical student at the Wayne State University in Detroit, who after a preliminary hearing launched his own personal *jihad* against the people of the United States. Law enforcement officers arrested him in a city park with camouflage paint on his face and carrying a loaded AK-47.

Still another, Youssef Megahed and Ahmed Mohamed, both students at the University of South Florida, were stopped for speeding near the Goose Creek weapons base. Officers discovered a pipe bomb in the car.

"Terror experts say these and other cases illustrate an emerging threat from homegrown terrorists, people who have been radicalized by extreme Muslim doctrine within the U.S.," Fox reported.

Do you want to know how I would solve that problem?

Give them a camel apiece and send them all home!

Although there have been no major Al-Qaeda attacks on our land since 9/11, there have been numerous terrorist incidents.

Here in the United States an undercover survey of more than one hundred Islamic mosques and schools revealed intense radicalism and the alarming discovery that many of the mosques preach anti-West extremism.[7]

"Frank Gaffney, a former Pentagon official who runs the (private) Center for Security Policy, says the results of the survey have not yet been published," WorldNetDaily reported. "But he confirmed that 'the vast majority' are inciting insurrection and *jihad* through sermons by Saudi-trained *imams* and anti-Western literature, videos and textbooks."

According to Gaffney, preliminary findings reveal the *imams* are preaching that all non-Muslims, particularly Jews, are infidels and that *jihad* or holy war is a Muslim's duty. However, suicide bombers – such as those who attacked the Twin Towers on 9/11 – and other martyrs deserve the highest praise. They also teach that an "Islamic caliphate should one day encompass the U.S."

"Though not all mosques in America are radicalized, many have tended to serve as safe havens and meeting points for Islamic terrorist groups," WorldNetDaily reported. "Experts say there are at least 40 episodes of extremists and terrorists being connected to mosques in the past decade alone."

The report also noted that some of the 9/11 hijackers received help from the Dar-al-Hijrah Islamic Center in the Washington, D.C. area. It is identified as one of the mosques at the center of terrorist activity and extremism.

"It was founded and is currently run by leaders of the Muslim Brotherhood," WorldNetDaily said. "*Imams* there preach what is called '*jihad qital*,' which means physical *jihad*, and incite violence and hatred against the U.S."

According to the investigators, their goal is to turn the United States into an Islamic country governed by *sharia* (Islamic religious) law.

To my knowledge, no government agency has investigated the alleged anti-American activities inside the mosques for fear of alienating the thousands of Muslims who live in this land.

Meanwhile, a charter school in Inver Grove Heights, Minnesota "is named after a Muslim warlord, shares the address of the Muslim American Society of Minnesota, is led by two *imams,* is composed almost exclusively of blacks, many Somalis, and has as its top goal to preserve 'our values.'"[8]

The school is supported by the taxpayers of Minnesota.

Robert Spencer, director of *Jihad* Watch and an Islamic history and theology scholar, asked, "Can you imagine a public school founded by two Christian ministers, and housed in the same building as a church? Add to that – in the same building – a prominent chapel. And let's say the students are required to fast during Lent, and attend Bible studies right after school. All with your tax dollars."

"Inconceivable? Sure," Spencer said. "If such a place existed, the ACLU lawyers would descend on it like locusts. It would be shut down before you could say 'separation of church and state,' to the accompaniment of the *New York Times* and *Washington Post* editorials full of indignant foreboding, warning darkly about the growing influence of the Religious Right in America."

But the Islamic charter school has drawn no protests.

In August of 2007, Homeland Security operatives were informed by British authorities that they had uncovered a terrorist plot that would have killed thousands of innocent civilians on United States-bound flights from London, according to ABC News. It would have been the most destructive terrorist attack since 9/11.

"Terrorists who had planned to detonate gel-based explosives... would have achieved mass devastation, according to new information from (former) Homeland Security Secretary Michael Chertoff in an exclusive interview with ABC News," Pierre Thomas reported.

Chertoff said the terrorist attack would have been more severe than 9/11.

"If they had succeeded in bringing liquid explosives on seven or eight aircraft, there would have been thousands of lives lost and an enormous economic impact with devastating consequences for international air travel," Chertoff said.

According to ABC, the explosives can be made from components purchased at a drugstore or supermarket.

"There's no question that they (the terrorists) had given a lot of thought to how they might smuggle containers with liquid explosives onto airplanes," Chertoff said. "Without getting into things that are still classified, they obviously paid attention to the ways in which they thought they might be able to disguise these explosives as very innocent types of everyday articles."

Eight men went on trial in March of 2008 in Woolwich Crown Court for conspiracy to commit murder in connection with the planned terrorist attacks on several airliners bound for the United States from London.

"What these men intended was a violent and deadly statement of intent which would have a truly global impact," said Peter Wright, the prosecutor. "These men were actively engaged in a deadly plan designed to bring about what would have been, had they been successful, a civilian death toll from an act of terrorism on an almost unprecedented scale."[9]

The terrorists had targeted United Airlines, Air Canada

and American Airlines flights to Toronto, Montreal, San Francisco, Washington, D. C. and New York City.

"Plastic Oasis and Lucozade bottles were to be used by the plotters to make their liquid bombs," Charlotte Gill and Sam Greenhill reported for the *London Daily Mail*. "A hypodermic syringe would be inserted into the base to draw out the drink and the bomb mixture would be injected in its place… Bulbs and wires would connect the bomb mixture with disposable cameras to trigger a charge to set it off."

"The means by which they intended to inflict heavy casualties upon ordinary civilians was by the carrying out of a series of coordinated and deadly explosions," the prosecutor said. "These men were, we say, indifferent to the carnage that was likely to ensue if their plans were successful."[10]

Anti-terrorist police officers were said to have found one of the terrorist's diary that spelled out the plans for the attack.

According to the diary, gel-like explosives made to look like food coloring and mouthwash were to be smuggled on board the planes.

"There they would be hooked up to homemade detonators powered by tiny camera batteries and set off to cause mid-air carnage, the court heard," Gill and Greenhill reported. "The main ingredient of the homemade bombs was said to be hydrogen peroxide, commonly used as hair bleach and easily available on the high street, mixed with other chemicals…"

Syndicated columnist Cal Thomas has asked the question: "How much longer should we allow people from certain lands, with certain beliefs to come to Britain and America and build their mosques, teach hate, and plot to kill us?"

Ibrahim Hooper with the Council on American-Islamic Relations called Thomas an "Islamaphobe" and said that Muslim terrorists are a small minority of people "misusing the (Islamic) faith."[11]

Thomas called Hooper and his organization "disinformation people."

He said those groups are always trying to convince the American people there is no threat and the real threat is from men and women who warn others about the Islamic threat to this country and not "the people who are plotting to destroy us."

"The truth hurts," Thomas said. "I'm not making this stuff up... we gotta connect the dots. It's a little late after things begin to explode."

Meanwhile, Jan Markel, the founder of Olive Tree Ministries, a pro-Israel group from Minnesota, praised Thomas for taking a stand on Muslim terrorism.

Markel said she has not seen a rally by Muslims living in America denouncing terrorism.

"I don't see it happening," she said.

In other developments, British lawyer Paul Diamond, who represents various religious liberty cases, says the people of the United States and Great Britain should join together to "preserve the civilized world."[12]

Diamond says that although Muslims and homosexuals appear to be opposed on various issues, they are working together to push Christians out of the public square. He adds that both groups are fearful of the Judeo-Christian worldview because it is "attractive to the common man, used to be in power, and can be in power again."

"So rather than have an open debate and let the British or American people choose, the Christians have to fight by the Marquis of Queensbury rules... [with] one hand tied behind their back, so to speak," says Diamond.

The attorney says that the British and American Christians must work together to oppose what he calls "dark and primitive forces" that are the proponents of radical Islam.

"Diamond's recent clients include a British Airways employee prohibited from wearing a cross at work, a Christian teenager who was barred from wearing a purity ring at school, and a Christian magistrate desiring to be exempted from cases where he may have to put a child in a same-sex household," Jim Brown, writing for *OneNewsNow*, said.

Former Prime Minister John Howard of Australia is not in favor of introducing Islamic *sharia* law into Australia and said Muslims who don't like it should get out of the country.

Speaking at the neo-conservative American Enterprise Institute in Washington, D.C., he also strongly disagreed with the Rev. Rowan Williams, the Archbishop of Cantebury, who said earlier that the introduction of *sharia* or Islamic law into Great Britain was unavoidable. Reverend Williams is the spiritual leader of the world's Anglicans.[13]

"Mr. Howard dismissed the idea that *sharia* law can be introduced into Western societies, saying it is fundamental to the unity and purpose of a democratic nation state that there be only one body of law to which all are accountable," said Graeme Dobell reporting for the Australian Broadcasting Company. "The former Prime Minister says Western societies should not think they can trade away some of their values to gain immunity from terrorists or respect from noisy minorities."

Reverend Williams, during an interview with the British Broadcasting Company, called for a "constructive accommodation" with *sharia* law.

"Asked if the adoption of *sharia* law was necessary for community cohesion, Williams said, 'It seems unavoidable,'" BBC reported.

However, Prime Minister Gordon Brown responded to the archbishop by saying "*sharia* law cannot be used as a justification for committing breaches of English law, nor can the principle of *sharia* law be used in a civilian court."

"The prime minister is clear that in Britain, British laws based on British values will apply," BBC reported.

There are 1.8-million Muslims living in Great Britain and more arriving every day seeking citizenship.

"The issue of integrating Britain's... Muslims has been widely debated since July 2005, when four British Islamists carried out suicide bombings on London's transport network, killing 52 people," BBC reported.

Meanwhile, Howard told the members of the Enterprise Institute that radical Islam believes "that there is a soft underbelly of cultural self-doubt in certain Western societies."

"There are too many in our midst who think, deep down, that it is really 'our fault' and if only we entered into some kind of federal cultural compact, with our critics, the challenges would disappear," Howard said. "Perhaps it was this sentiment which led the Archbishop of Canterbury to make the extraordinary comment several weeks ago, that in Britain some accommodation with aspects of *Sharia* was inevitable."

Meanwhile, another cleric says that the collapse of Christianity is wrecking British society and that Islam is filling the moral vacuum.

Bishop Michael Nazir-Ali said the collapse has destroyed family life and left Great Britain virtually defenseless against the rise of radical Islam.

The Reverend Nazir-Ali, Bishop of Rochester, said that England is mired in a doctrine of "endless self-indulgence."

The Pakistani-born bishop said the downfall of Christianity in England began with the "social and sexual revolution" of the 1960s.[14]

Chapter Thirteen
The Disintegrating Halls of Ivy

There have also been significant changes in the colleges and universities of this land as most state-supported schools have strayed far away from the values of the founders of the erstwhile magnificent pillars of learning. Today's classrooms in the once-hallowed "Halls of Ivy" are a potpourri of liberal philosophy where some secular progressive professors use the classrooms as training grounds for the teaching of moral relativism.

Even the most casual observer recognizes there is hard evidence that moral relativism has placed our schools in peril and it is spreading like a virus across academic America.

Jim Nelson Black has written a book entitled *Freefall of the American University: How Our Colleges Are Corrupting the Minds and Morals of the Next Generation.* Black, who holds an earned doctor of philosophy degree, is the founder of the Sentinel Research Associates of Washington, D.C.

According to WorldNetDaily, the book carefully documents how today's universities are indoctrinating students with liberal/progressive agendas.

"Something very disturbing is happening in colleges all across the country," Black says. "Instead of being educational institutions designed to encourage the free discussion of ideas, universities have become prisons of propaganda indoctrinating students with politically correct (and often morally repugnant) ideas about American life and culture."

Black's book clearly exposes today's liberal bias in education and how colleges are promoting a left-wing philosophy of sexuality, politics and lifestyles.

Here are some of the most amazing events and developments in our institutions of higher learning.

A University of Maine student says one of her professors offered extra credit for anyone who would burn the American flag or the U. S. Constitution or would be arrested for protesting on behalf of free speech.

Sophomore Rebecca McDade said the offer was made by Professor Paul Grosswiler in his *History of Mass Communications* class. McDade, who was upset over the professor's comment, dropped the class.[1]

"I was offended... the flag and the Constitution are really important symbols to me because of my family background," she said.[2]

The professor said he was misunderstood, a typical response from leftist university professors when they are challenged.

"When you expose leftist abuses... I'm sure that the administration, like most administrations we deal with, is not happy when leftist abuses come to life," says Morton Blackwell of the Virginia-based Leadership Institute that is assisting McDade.

Blackwell says his organization wants the University to adopt a Students Bill of Rights to protect the students from professors who treat and grade them based on their religious or political beliefs.

McDade said when the professor mentioned the flag-burning, the fifty students in the class were quiet and some questioned whether he was serious.

Kathleen Dame, another student in the class, said at first she thought the professor was joking until he explained that burning the flag was not illegal.

A professor at the Suffolk Community College in New

York demands that his students acknowledge the possibility that God does not exist in order to take his philosophy class which is required for graduation.

Gina DeLuca, a student in the class, was labeled "closed-minded" by the professor who lowered her grades because of her beliefs.

The American Center for Law and Justice (ACLJ) presented a demand letter to university officials to stop the professor's religious persecution, a prelude to a federal lawsuit.[3]

"This is another terrible example of how some in the academic world believe it's acceptable to violate the First Amendment rights afforded to all students, especially students who hold Christian beliefs," says Jay Sekulow, chief counsel for ACLJ.

Sekulow said the professor's actions reveal his hostility toward religion.

According to Sekulow, DeLuca holds a 3.9 grade point average and received good grades in her philosophy class until the professor learned of her Christian faith.

"The grades she received on class assignments dropped significantly once God and religion became prominent topics of the class discussion and her refusal to compromise her Christian faith became apparent," Sekulow said.

He said that the professor also called DeLuca "closed-minded," "uncritical," and "blinded by belief."

"While a college professor may encourage students to be informed about viewpoints and arguments that differ from their own, it is inappropriate – and unconstitutional – for a public college professor to make passing a required course (and thus graduation) contingent upon a student's willingness to express agreement with philosophical viewpoints that conflict with her religious beliefs," he said.

The University of Delaware planned to initiate a mandatory course of indoctrination that branded all white students as "racists." The program would have required about 7,000 residence hall students to take the course. The university planned to provide "treatment" for any of the so-called "incorrect attitudes" regarding gender, religion, culture and sexuality, according to a civil rights group concerned about the program.

The University would have required students to adhere to the school's worldview on such issues as politics, race, sexuality, sociology, moral philosophy and environmentalism.

"Somehow, the University of Delaware seems terrifyingly unaware that a state-sponsored institution of higher education in the United States does not have the legal right to engage in a program of systematic thought reform," said Samantha Harris of the Foundation for Individual Rights in Education. "The First Amendment protects the right to freedom of conscience – the right to keep our innermost thoughts free from governmental intrusion. It also protects the right to be free from compelled speech."

Harris further pointed out that the shocking program of ideological reeducation which the school defined as a "treatment" for students' incorrect attitudes and beliefs "is nothing less than Orwellian."

Greg Lukianoff of the Foundation said that according to university training documents:

"A racist is one who is both privileged and socialized on the basis of race by a white supremacist (racist) system. The term applies to all white people (of European descent) living in the United States regardless of class, gender, religion, culture or sexuality."

Lukianoff said the fact that such a program ever existed in a public university is stunning.[4]

Walter E. Williams, an economist at George Mason University and also a black scholar, said the University's definition of "racist" is, in essence, reverse racism.

"This gem of wisdom suggests that by virtue of birth

alone, not conduct, if you're white, you're a racist," he said in a column.

Because of the effective work of the foundation and thousands of parents, the University dropped the indoctrination program.

The University of Vermont at Burlington has a beautiful new student center but if you're looking for the men's or women's rooms you're out of luck.

All they have is gender-neutral bathrooms primarily for transgendered students and visitors to the campus.

The Associated Press reported that there are signs outside the new facilities that read "gender neutral restroom."

Annie Stevens, a vice president for student life, said the gender-friendly rest rooms will meet all the people's needs.[5]

Gender-neutral restrooms on some college campuses are part of a secular progressive trend that also includes coed dorm rooms on various campuses where young men and women can choose to share a room together.

"At least two dozen schools, including Brown University, the University of Pennsylvania, Oberlin College, Clark University and the California Institute of Technology, allow some or all students to share a room with anyone they choose – including someone of the opposite sex," Michelle R. Smith reported for the Associated Press.

More schools, including Stanford University, are also expected to endorse coed dorm rooms.

The historic College of William and Mary was established in 1693 and is the second oldest educational institution in the nation after Harvard. The Rev. Dr. James Blair, an Episcopalian clergyman, founded the school in 1693 based on Christian principles.

Wren Chapel is one of the landmarks on the campus where, until 2006, the 100-year-old Wren Cross was displayed on its altar table.

President Gene Nichol ordered the cross removed from the altar because he thought it might offend some of the visitors to the campus.[6]

However, the board of regents overruled Nichol and ordered the cross returned to the chapel. But rather than displaying the cross on the altar, they decided to place it in a glass case as part of the historical heritage of the college.[7]

They made the decision after a wealthy benefactor of the school threatened to withdraw a $12-million gift to his *alma mater*.

But get this, President Nichol approved plans for a Sex Workers' Art Show on the campus that would feature topless dancers and prostitutes.

Nichol later resigned as president of the school.

Ward Churchill was a professor at the University of Colorado at Boulder for 17 years until he was fired in 2007.

Churchill received a degree of notoriety when he wrote an essay in which he referred to the terrorist attacks on 9/11 as "chickens coming home to roost," and said the 2,974 victims deserved to die.

After the attack, he said it was a reprisal against America's unjust foreign policy in the Middle East.

"One of the things I've suggested is that it may be that more 9/11s are necessary…" Churchill said. "I want the state gone: transform the situation to U.S. out of North America. U. S. off the planet. Out of existence altogether."

Through the years, Churchill's academic career was characterized by his radical beliefs that the United States has been responsible for genocide, the extermination of large groups of people. He once said that this country is worse than Nazi Germany and that beginning in 1492 it "unleashed a process of conquest and colonization unparalleled in the history of humanity."[8] He also said Christopher Columbus was as bad as the German Gestapo chief

Heinrich Himmler.[9]

Churchill's former students said he actually equated the United States with Germany's Third Reich during his class on the *American Holocaust*.[10]

Bill O'Reilly of the Fox News Channel noted that the so-called "free speech crowd" rushed to defend Churchill.

"The ACLU urged CU (Colorado University) not to fire the man, and a bunch of other phonies screamed 'academic freedom,'" O'Reilly said. "This is a far-left advocacy group with no interest in anybody's speech it doesn't like."

I was absolutely stunned when I learned that Lee Bollinger, the president of Columbia University in New York, had invited Iranian President Mahmoud Ahmadinejad to speak at the school.

There were several reasons why, in my opinion, it was a bad idea.

First, Ahmadinejad's government is supplying weapons of war to the Al Qaeda terrorists to be used against American troops in Iraq and Afghanistan.

Second, he has denied that the "holocaust" ever occurred.

Third, he believes that Israel should be wiped out.

Fourth, he hates America and called for a global *jihad* to destroy us.

"Today, the time for the fall of the satanic power of the United States has come and the countdown to the annihilation of the emperor of power and wealth has started," he said in a speech in early June of 2008.

It also appeared to me that the University's invitation to Ahmadinejad was a slap in the face of the 1.4-million Jewish people living in the metropolitan New York City area.

But Bollinger defended his decision to have the controversial speaker address the student body.

According to the Associated Press, at least two New York newspapers questioned the visit. *The New York Daily*

News carried a headline that read: "The Evil Has Landed."
The New York Post called him a "madman" and a "guest of
dishonor."

The University of Saint Thomas, a Catholic school in
Minnesota, denied permission for columnist Star Parker,
an African-American author and TV personality, to appear
at the school during her college speaking tour.[11]

"How could this be happening?" she asked. "A Catholic
university, which has played host to talks from left-wing
Al Franken and a transgendered woman named Debra
Davis, nixes a presentation on abortion from a conservative
black woman?"

After hundreds complained, administrators at Saint
Thomas relented and placed her back on the school calendar.

"I've been speaking on university campuses for years –
more than 150 of them," she said. "I know firsthand their
left-leaning bent and what a conservative has to deal with
walking into the belly of the liberal beast... But this incident had a more bitter tinge than usual. This was a Catholic university, and my topic was abortion."

Parker said her rejection by Saint Thomas ironically
came on the day that Pope Benedict XVI – who has called
abortion "today's greatest injustice" – was to arrive in the
United States.

"Pope Benedict... has talked about the 'dictatorship of
relativism,'" she said. "This accurately captures what we're
dealing with on our campuses."

What about the hallowed halls of Harvard?

The administration of the elite university has given special "exercise" privileges to Muslim women. By so doing,
Harvard has joined a growing list of colleges and universities giving concessions to Muslims.[12]

The Associated Press reported that Harvard officials

told all male students to stay away from the gyms during certain hours of the week to give the Muslim women their privacy during exercise periods.

Jan Markell, founder and director of Olive Tree Ministries, says the Muslim women are given privileges that no other women's groups receive.

"It's the principle – and the principle is that political correctness is appropriated only to the Muslim community in America… They always get their way," Markell said.

Markell says she wonders how the Harvard officials would respond if a Christian or Jewish group would ask for the same privilege.

"I think if almost any group, particularly Christians, asked for such special rights or conditions… we would be laughed off the planet by just about everybody," Markell said.

A Harvard spokesman noted that the university also provides designated areas for Muslim and Hindu students to pray.

Angry students at Washington University in St. Louis, Missouri, protested the school's decision to award an honorary degree to Phyllis Schlafly, founder of the conservative Eagle Forum and one of the leading opponents of the Equal Rights Amendment back in the 1970s.

Students demonstrated in front of the home of Mark Wrighton, the school chancellor, chanting and carrying signs asking him to rescind the degree.[13]

Wrighton refused and said Schlafly's life and work "have sparked debate and had a broad impact on American life."

During the presentation of the award, hundreds of graduating students and several faculty members on the platform turned their backs as Schlafly was honored, according to the *St. Louis Post-Disptatch*.

She just smiled.

Later she said the graduating students' actions were "juvenile" and they were "raining on their own parade."[14]

She previously called the demonstrators "a bunch of losers."

The 83-year-old Schlafly holds both a bachelor's and law degree from the institution.

I don't carry a copy of the Constitution around in my hip pocket, but the last time I read it there was a freedom of speech clause in the First Amendment. Sometimes I wonder why there is such vitriol from college students against a great leader like Phyllis Schlafly.

Certain students and faculty members at Washington University – and others all across this land – want freedom of speech for themselves to promote every kind of left-wing cause imaginable, but are opposed to Schlafly having the same constitutional privilege.

That's demagoguery!

If those faculty and student demonstrators are so convinced they are right in their secular progressive worldview, why would they be afraid of an 86-year-old woman such as Phyllis Schlafly?

Ironically, the freedom of speech clause in the First Amendment also allows all of us to make fools out of ourselves.

Lesbians at the elite Smith College in Northampton, Massachusetts, rioted during a speech by Ryan Sorba on *The Born-Gay Hoax*.

"We're here, we're queer, get used to it!" the lesbians shouted as they beat on pots and pans. Police and security officers stood by and did nothing.

The lesbians also flooded the speaker's platform and began dancing and clapping in front of Sorba.

School officials decided to stop Sorba's presentation.

Alas! I wonder if the officials even thought about stopping the lesbians?

This event, like so many others, raises an important question? If the lesbians are so sure they are right – that they are born with a lesbian gene – why are they so afraid of a different opinion? Yet shouting down speakers with opposing views has become a standard practice among lesbians/homosexuals.

"The Smith Republican Club had invited Sorba to speak on his soon-to-be released book debunking the claim that homosexual orientation is genetically determined, and hence, worthy of social acceptance and legal protection," *LifeSiteNews* reported of the incident. "Homosexual activists routinely use the notion of the 'gay gene' to argue that any opposition to their lifestyle is unjust discrimination."

Lloyd Jacobs, the president of the University of Toledo in Ohio, fired vice president Crystal Dixon, a black woman, because of her beliefs concerning homosexuality.

Dixon, head of the human resources department at the school, wrote a letter to the *Toledo Free Press* newspaper after an editorial writer complained that homosexuals were not being treated fairly at the school and apparently referred to them as civil rights victims.

She disagreed and took "great umbrage at the notion that those choosing the homosexual lifestyle are 'civil rights victims.'"

"Here's why. I cannot wake up tomorrow and not be a black woman," Dixon said in the letter.

However, like many others, Jacobs believes the homosexuals can change if they choose to do so.

When Jacobs read the letter to the editor, he immediately suspended Dixon and condemned her opinion.[15]

Several days later Jacobs convened a "pre-disciplinary" meeting concerning her beliefs and determined she no longer had the ability to carry out her duties and responsibilities.

I've wondered what one's opinion of homosexuality has

to do with his/her ability to do a good job in a chosen profession.

"After reviewing the relevant evidence... it is my determination that there is just cause to terminate your employment with the University of Toledo," Jacobs wrote, condemning her opinions as "in direct contradiction" to his policies and procedures.

Brian Rooney a lawyer with the Thomas More Law Center said it is obvious that Dixon's First Amendment rights were violated.

"Then there's the equal protection clause," Rooney said. "We know of instances of administrators similarly situated who have written opinion pieces saying great things about homosexuality. Nothing has happened to them."

He said his law firm will also explore other issues because she "clearly was fired for her religious beliefs."

"The Supreme Court has stated clearly, as a matter of constitutional law, if you're a public employee you can speak as an individual on matters of public concern," Rooney said. "It's protected speech."

One of the tragedies in higher education today is the persecution of Christians, like Dixon, who hold a Christian worldview, the antithesis of moral relativism.

I want to tell you the story of professor-emeritus Dean Kenyon at the San Francisco State University in California who has been persecuted for years because of his beliefs.

Through the years Dr. Kenyon was recognized as one of America's leading authorities on chemical evolution theory and the origin of life. He wrote a book entitled *Biochemical Predestination* that was popular in the biology departments of various universities across America and around the world, even in the Soviet Union.

The controversy began when John Hafernik, the chairman of the biology department at the University, ordered Kenyon to teach only evolution in his introductory biology classes.

Kenyon for years, during his lectures to students on biological origins, exposed them to both the theory of evolution and evidence against that theory.[16]

"He also discussed the philosophical and political controversies raised by the issue and his own view that living systems display evidence of intelligent design – a view not incompatible with some forms of evolutionary thinking," said Stephen C. Meyer writing for the *Access Research Network*.

After receiving instructions to teach only evolution theory, Kenyon wrote Dean James Kelley and asked if he was "forbidden to mention to students that there are important disputes among scientists about whether or not chemical evolution could have taken place on the ancient earth."[17]

"Mr. Kelley replied by insisting that Mr. Kenyon teach the dominant scientific view not the religious view of special creation or a young earth," Mayer said.

Kenyon replied that he always presented the so-called "dominant scientific view" to the students but also explained to them that some biologists believe there is evidence for intelligent design.

He never received a reply but Kelley removed him from teaching any biology courses and assigned him to labs.

Kenyon, who received his Ph.D. in biophysics from Stanford and did post-doctoral studies at the University of California, Berkeley, and at Oxford in England, began to question some of his own ideas back in the 1970s.

"Experiments (some performed by... Kenyon himself) increasingly contradicted the dominant view in his field," Mayer said. "Laboratory work suggested that simple chemicals do not arrange themselves into complex information-bearing molecules such as DNA – without, that is, 'guidance' from human experimenters."

The idea that Kenyon's view is based on "religion" and the materialistic view is "scientific" reveals a double standard among the science professors within an area of science that is "notorious for its philosophical overtones," Mayer said.

When the University's Academic Freedom Committee reviewed Kenyon's case, they ruled decisively in his favor.

"The committee determined that, according to University guidelines, a clear breach of academic freedom had occurred," Mayer said.

Dean Kelley disagreed with the committee's recommendation, but reinstated him to his classes.

Professor Guillermo Gonzalez, a respected astronomer with stellar academic credentials at Iowa State University, was denied tenure because of his belief in intelligent design.

Gonzalez believes university officials took the action because of a book he wrote entitled *The Privileged Planet* which sets forth the premise that there is evidence of design in the universe.

"Dr. Gonzalez's case has truly distressing implications for academic freedom in colleges and universities across the country, especially in science departments," said Ken Conner writing for *TowhHall.com*. Conner, who is chairman of the Center for a Just society in Washington, D.C., is also a nationally recognized trial lawyer.

The professor earned his Ph.D. in astronomy from the University of Washington and is also the author of 68 scientific papers.

"These papers have been featured in some of the world's most respected scientific journals, including *Science* and *Nature*," Conner said. "Dr. Gonzalez has also co-authored a college-level text book entitled *Observational Astronomy*, which was published by Cambridge Press."

So why did the University withhold tenure from him?

"So far as anyone can tell, this rejection had little to do with his scientific research, and everything to do with the fact that... Gonzalez believes the scientific evidence points to the idea of an intelligent designer," Conner said.

It's interesting to note that the professor does not teach intelligent design in any of his astronomy classes and none of his scientific papers deal with the subject.

"Nevertheless, simply because Gonzalez holds the view that there is intelligence behind the universe, and has written a book presenting scientific evidence for this fact, he is considered unsuitable at Iowa State," Conner said.

Conner says the University's unfair treatment of Gonzalez raises several questions:

"What is the state of academic freedom when well qualified candidates are rejected simply because they see God's fingerprints on the cosmos?... Aren't universities supposed to foster an atmosphere that allows for robust discussion and freedom of thought?"

He adds that even though many science departments in our nation's universities are doing everything they can to stop professors from teaching intelligent design, more and more professors are embracing the concept.

"The fact that these (evolutionary) scientists, who are supposedly open to following the evidence wherever it leads, have resorted to blatant discrimination to avoid having this conversation speaks volumes about the weakness of their position," Conner said. "They realize their arguments are not sufficient to defeat the intelligent design movement and they must, therefore, shut their opponents out of the conversation."

Professor Gonzalez has appealed his case and Conner believes the decision not to grant him tenure will be reversed.

"Nevertheless, what happened to Dr. Gonzalez is a reflection of the growing strength of the intelligent design movement, not its weakness," Conner said.

Today's college and university students hear so much

progressive propaganda they never learn anything about the classical philosophers or how Christianity helped liberate blacks and supported women's rights in America, rights that are denied in many other places in the world.

So said Phyllis Schlafly in *The Phyllis Schlafly Report* in April, 2006.

"Most college students will not hear about the role Ronald Reagan... played in ending the Cold War..." she said. "Students in literature classes will read the latest novels about alienation, oppression, suicide and sex, but they won't encounter the novels of James Fenimore Cooper, Joseph Conrad, or Dostoevsky."

There is abject censorship of the great accomplishments of Western civilization and the great literary works in the English language in our institutions of higher learning. And "that is why it is that people want to come here from all over the world," Schlafly said.

According to Schlafly, Young America's Foundation (YAF) surveyed courses in several prestigious colleges and found that some very interesting and bizarre subjects are in the curricula. The YAF formed a list it calls the "Dirty Dozen."

"At Harvard, students can take a course in *Marxist Concepts of Racism*," Schlafly said the report revealed. "Karl Marx didn't say much about racism, but that course is an excuse for a left-wing professor to talk about racism."

The report said that at Johns Hopkins there is a course on *Sex, Drugs and Rock 'n' Roll in Ancient Egypt*. The course includes a visual presentation of Egyptian women drunk and vomiting.

Think about it! Queen Nefertiti rockin' around the clock with one of the pharaohs and then barfing all over the palace floor!

I'm sure that will greatly enhance the students' college careers.

Princeton offers a course entitled *The Cultural Production of Early Modern Women* which deals with cross-dressing and prostitution in England, France, Italy and Spain in the 16th and 17th centuries.

"Marxism, homosexuality and racial theory are current themes in college classes," Schlafly said. "Here is another shocking course title: *The Unbearable Whiteness of Barbie: Race and Popular Culture in the United States* in which students are taught that so-called 'scientific racism' was used to market the Barbie doll."

Students at UCLA can enroll in a course called *The Psychology of the Lesbian Experience* and Swarthmore College has a course in *Lesbian Novels Since World War II.*

Others on the YAF's "Dirty Dozen" list include: Amherst College's *Taking Marx Seriously: Should Marx Be Given Yet Another Chance?;* Brown University's *Black Lavender: A Study of Black Gay and Lesbian Play;* and *Ancient Greek to Modern Gay Sexuality* at the University of Michigan.

"One reason college tuition is so high is that colleges are paying high-priced professors to teach this radical nonsense," Schlafly said.

Are our colleges and universities discriminating against American young people? Schlafly believes they are.

"The University of Illinois seems to favor admitting foreign students rather than Americans," Schlafly said. "Students from India, Pakistan, China, Korea and dozens of other countries make up 12% of the student body. The University trustees are currently planning to increase foreign students to 25%... and at the same time reduce the freshman class by 1,000 students."

Schlafly points out that will mean that hundreds of qualified students from Illinois will not be able to attend the University in their own state because of the policy of admitting foreign students.

The chancellor defends the plan by saying the school has an obligation to students regardless of their nationality.

"Sorry, Mr. Chancellor, you've got your priorities all

backwards," Schlafly said. "Your first responsibility is to the taxpayers of Illinois. There is nothing wrong with admitting some foreign students, but it is just plain wrong to admit such a high percentage of foreigners over qualified Illinois students."

The great numbers of foreign students attending our colleges and universities also create another problem: the universities hire large numbers of the foreign student graduates as teachers in math and science courses, according to Schlafly.

"One of the most frequent complaints I hear on college campuses is that so many math and science instructors don't speak intelligible English..." she said. "State universities have the obligation to provide math and science instructors whose native language is English. The way to assure this is admit more Americans rather than foreigners to get their college degrees in math and science."

In her report, Schlafly recommends a book that all college students should read. It is David Horowitz's *The Professors: The 101 Most Dangerous Academics in America.* She says the book exposes how professors are imposing their own brand of radicalism on the students. Horowitz is a conservative writer and activist.

"When asked what's the typical ideological profile of 101 professors in his book, Horowitz replied: 'A professor who believes the (Islamic) terrorists are freedom fighters and America is the Great Satan. They all believe in some version of Marxism, though they call it feminism, post-structuralism, post-modernism. They believe private property is evil... America is an oppressive, racist, homophobic, imperialist society and the reason we are being attacked (by *jihadists*) is because of what we do to other people. Jews are monsters who have turned Palestinians into suicide bombers because Palestinians have no other choice. That's the fundamental creed of the tenured radical."

Horowitz says his book will scare any parent who has a

child enrolled in a college or university.

Our great educational institutions have graduated leaders in the world of math and science, literature, business and economics and other disciplines. These were some of the people who helped make American the greatest country on earth.

Who could ever have imagined that, after the magnificent rise of these institutions, we would witness their rapid fall, the fall we have witnessed with our own eyes since the 1960s?

Chapter Fourteen
Oprah's New Religion

The cataclysmic paradigm shift taking place in today's culture also has given birth to a new religion, one that is spreading across America like wildfire. It's a resurgence of New Age teaching and Oprah Winfrey is one of the leaders. She uses her vast financial empire, popular television and radio programs and publications such as *O Magazine* and Oprah's Book Club to attract millions.

What's this new religion all about?

It is a syncretism of religious beliefs from Eastern mysticism and Christianity and Oprah has become the exemplar and prophetess.

Back in the 1980s while serving as senior editor of Huntington House Publishing Company, I edited the first book exposing the New Age Movement, a book entitled *The Hidden Dangers of the Rainbow* by Constance Cumbey, a brilliant attorney from Michigan. Later I worked with a young author by the name of Randall Baer who had been a well-known leader in the movement before his conversion to Christianity. His book was *Inside the New Age Nightmare.*

After we at Huntington House opened the door, dozens of authors wrote books or so-called exposes on the movement. However, with the passing of time, I came to believe the movement had died a slow but well-deserved death.

Oh, there were appearances by movie stars like Shirley MacLaine on Larry King Live to discuss reincarnation (a major tenet of New Age thought) and her previous lives but most of us thought her experiences were more fantasy than fact.

Some of us even laughed when she said in a previous life she had been the lover of King Charlemagne of France, a *geisha* in Japan and a model for famed artist Toulouse-Lautrec.

Or when Linda Gray, the neurotic Sue Ellen on the hit TV program *Dallas,* was the subject of a brief *Remembering Your Spirit* segment on Oprah's program. There was Linda in an Indian teepee somewhere in New Mexico sitting cross-legged before an open fire waving smoke all around in an attempt to receive New Age enlightenment.[1]

Then there is Linda Evans, the beautiful Audra on *The Big Valley* TV show and Krystle on *Dynasty* who claims to have her own personal channel, a being from another dimension who speaks to her and helps her along the New Age pathway.

But we are no longer laughing for the revival of the New Age religion, under the leadership of Winfrey and high-level *gurus,* presents a serious challenge to orthodox Christianity in this land.

I wonder if this flawed new age religion has emerged to fill a vacuum in the traditional church in what some have labeled "the post-Christian era."

Let's take a brief look at the New Age Movement to make sure we understand what it's all about.

Author Marilyn Ferguson, a moving force back in the 1980s, wrote a seminal book entitled *The Aquarian Conspiracy.* She prophesied radical change or a "new age" would become a powerful spirituality in the world. Her book became the New Age Bible.

Here's how she described the movement: "A leaderless but powerful network to bring about radical change in the United States. Its leaders have broken with key elements of Western thought and they may even have broken continuity with history."[2]

New Age literature – though broad and encompassing

and often quite ethereal – reveals the movement combines modern occult traditions replete with psychics, familiar spirits and medium channels. It also embraces Eastern and Western mystical religions and goddess worship.

In a sense, it is modern sorcery that is strictly prohibited in God's Word.

Ferguson gave us some insight into the types of people who have joined the new religion.

"The Aquarian conspirators are... school teachers and office workers, famous scientists, government officials and lawmakers, artists and millionaires, taxi drivers and celebrities, leaders in medicine, education, law, psychology," she wrote. "Some are open in their advocacy. Others are quiet about their involvement, believing they can be more effective with ideas that all too often have been misunderstood. "

Back in the 1980s I became acquainted with Randall Baer, a New Age teacher, author and activist. He wrote two successful books on the subject of crystal power published by Harper and Row. He also built a large teaching and research facility in New Mexico.

Baer once told me he was convinced the New Age was the true path to enlightenment, love and peace.

"What once was considered to be only a passing phenomenon by some observers is continuing to grow, gain influence, and capture the minds and allegiance of millions..." he said back in the 80s.[3]

According to Baer, most people in the movement are deceived and have no idea of the underlying sinister dangers. "Brainwashing," he said, "blinds a person to the horrors that lurk beneath the surface of the New Age."

Baer told me he was so deep into the metaphysical he was writing best-sellers that had been given to him through "automatic writing" by his spiritual channel from another spiritual and physical dimension.

However, once while Baer was in a blissful trance, God

supernaturally removed the veil and suddenly he saw the evil spirits that were creating the illusion he was experiencing. He believed the vision was a warning from God and began to seek the true God and became a Christian. He wrote the *Nightmare* to shine a light on the darkness he once embraced.

He later denounced his previous books on crystal power and asked the publisher to discontinue publication. The publisher declined.

Commenting on the scope of the movement, Baer said, "What once was considered to be only a passing phenomenon by some observers is continuing to grow, gain influence and capture the minds and allegiance of millions..."

But Baer says the New Age Movement is nothing new at all.

"It always has been active throughout history in numerous and widespread Western occult traditions and Eastern mystical religions," he said. "Over the last three decades, however, enormous and unprecedented massive revival of the occult-based practices has been taking place..."[4]

He issued a warning, especially to Christians. He believed large numbers of people would be seduced by New Age promises of godhood, unlimited power and immortality through incarnation.[5]

"When the eternal truth of the Lord seems to a New Ager to be utterly false, and the falseness of the New Age seems to be absolutely true, therein lies an intricate portrait of seduction," he said.

Now, what about the Oprah connection?

USA Today reporter Ann Oldenburg wrote an article entitled *The Divine Miss Winfrey?*

"After two decades of searching for her authentic self – exploring New Age theories, giving away cars, trotting out fat, recommending good books and tackling countless issues from serious to frivolous – Oprah Winfrey has risen to a new level of *guru*," Oldenburg wrote.

Kathryn Lofton, a professor at Reed College in Portland, Oregon, has written two scholarly papers dealing with Oprah's religious beliefs and says, "She's a really hip and materialistic Mother Teresa" and points out that she "has emerged as a symbolic figurehead of spirituality."

"She's a moral monitor, using herself as the template against which she measures the decency of a nation," Lofton says.

Mark Jurkowitz writing in *The Phoenix* said, "She puts the (New Age) cult in pop culture."

In a sense, millions of women may have elevated her to a cult goddess status.

"Love her or loathe her, Winfrey has become proof that you can't be too rich, too thin or too committed to rising to your place in the world," Oldenburg said. "With 49 million viewers each week in the USA and more in the 122 countries to which the show is distributed, Winfrey reaches more people in a TV day than most preachers can hope to reach in a lifetime of sermons."

Religion writer Cathleen Falsani, writing for the *Chicago Sun-Times,* asked: "I wonder, has Oprah become America's pastor?"

Marcia Nelson in her book *The Gospel According to Oprah* describes Oprah as a spiritual leader and adds, "I've said to a number of people – she's today's Billy Graham."

Hence, large numbers of those people are turning to the Church of Oprah.

Nelson adds, "People who have no religion relate to her."

"Oprah is not only an effective communicator, but she is also an effective evangelist for Postmodernism, a world view that allows the individual to accept truth on his own terms as it relates to his experience…" said Russ Wise writing for the *Christian Information Network*. "As a result, she has become an international spiritual leader in her own right."[6]

Oprah has turned millions of her adoring fans over to the New Age gospel and has said on her program that it is a mistake to believe there is only one way to find God.

She once told her television audience – made up mostly of women – that she had a religious experience whereby she opened her mind to the "absolute indescribably huge-ness of that which we call God."

After that experience, she rejected rules, belief systems and doctrines.

Let's take a look at Oprah's new religious doctrines and how she is sharing her beliefs with millions in one of the most aggressive evangelistic movements in history as she has become the leading spokeswoman for the New Age.

Oprah gradually gravitated away from her Baptist heri-tage to embrace a broad range of New Age doctrines in-cluding the following:

- Heaven is not a location but refers to an inner realm of consciousness.
- Jesus on the cross was only an image that is in every man and woman.
- Each person's mind is a part of God's mind.
- God is in all things.
- Our holiness is our salvation.
- My salvation comes from me.
- There is no sin.
- It is a "pathetic error" to cling to the "Old Rugged Cross."
- The only message in Christ's crucifixion is that everyone can overcome the cross.
- "There couldn't possibly only be one way," she says. "There are millions of ways and many paths to what you call God."

- Each person has an inner guide.
- And dozens of other beliefs too numerous to mention.

Hence, in recent years she has adopted the New Age pantheistic worldview. Pantheism teaches that God is not the personal being of the Bible, but a universal force or energy and is in everything.

How did she come to believe these anti-Biblical and extra-Biblical teachings? Who were the spiritual mentors that influenced her metamorphosis? What was it in their message she found so captivating?

"These questions are not only appropriate to ask, but they are critical to our understanding of how Oprah became seduced by the perennial teachings of the occult," Wise said.

She had a lot of help from her spiritual mentors such as Deepak Chopra, Marianne Williamson, Jean Houston – a counselor to the Clinton White House – Eric Butterworth and a host of others. She also acknowledges she routinely consults an inner guide or the "divine" within her.

She also believes strongly in Hindu teaching that "truth is one, paths are many."

Therefore, a person may want to choose the Christian path or the Hindu for it makes little difference since they all lead to the same truth.

Through the years, Oprah's television programs were geared to entertain millions of people. Now that has changed as she believes there is a far more important "ministry" to help her millions of viewers recognize their own divinity.

"She has become the most prominent evangelist of popular culture regarding moral and spiritual issues and she is being heard," Wise said.

Various polls indicate that Americans are being greatly influenced by Oprah and the New Age guests she invites to be on her program.

She provides a platform for teachers to help her viewers find individual empowerment in their religious quest.

"Oprah has given numerous New Age personalities a platform to do exactly that, to offer her viewers a religion that offers empowerment rather than a religion that offers the individual a relationship with their creator." Wise said. "These self-styled spiritual leaders to whom Oprah gives a podium seduce her audience with their occultic message…"

According to Wise, the New Age/occultic influence of Oprah's spiritual mentors is without question. "Her interviews with each of them are examples of… the Master who enlightens the student and the student who offers devotion to the Master. Oprah has fully accepted her role as seeker and teacher of the self-empowering philosophy. She sees her role as the conduit between the enlightened and those in darkness."

All of the mentors seem to repeat the same *mantra:* there is a power or a force within each person that reveals the God-self or the innate divinity in each person, Wise explains.

Wise says one of Oprah's most influential mentors is Eric Butterworth, a former senior minister of the Unity Center in New York City, who also teaches a pantheistic view of God or that God is only a universal force or energy.

"Pantheism supports the idea that God, as a universal force, is in each individual regardless of the person's spiritual state or whether the individual believes in God or not," Wise said.

Butterworth taught Oprah that Jesus was merely a teacher of "Divine" truths who helped man recognize his own divinity.

Wise points out that Butterworth teaches that individuals find salvation through good works after many lifetimes (reincarnations) and therefore redemption is unnecessary. This point of view does not accept the belief that anyone can have a personal relationship with the Creator.

"Butterworth makes this point time and time again," Wise said. "His basic premise is that the power within each individual is God – the force or energy that permeates all of creation."

Wise believes Butterworth uses scripture to prove his personal beliefs.

"For example he states that, 'The message of the Gospels has been misunderstood. They have been made to appear to say that Jesus was really god taking the form of man... It fails to catch the real theme of His teaching: the truth of the Divinity of Man," he said.

Butterworth says in his book *Discover the Power Within You*, "We must study and emphatically reject our historical tendency to worship Jesus. When he becomes the object of our worship, He ceases to be the way-shower for our own self-realization and self-unfoldment."[7]

Marianne Williamson is another one of Oprah's mentors.

Williamson is a best-selling author of metaphysical books but perhaps is best known for *A Course in Miracles* which she has made popular. When she appeared on Oprah's show in 1992 she made the startling statement: "'There is only one begotten Son' doesn't mean that someone else was it, and we're not. It means we're all it. There's only one of us here."

Oprah's promotion of the *Course* made it an overnight sensation and Williamson a major player on the New Age scene.

But Warren Smith, a former follower of the New Age and author of *Reinventing Jesus Christ: The New Gospel*, said any Christian should recognize the false teaching of *A Course in Miracles*.

According to Smith, here are some of the false teachings: Lesson 29, "God is in everything I see." "Lesson 186: "The salvation of the world depends on me." Lesson 253: "My self is the ruler of the universe." Lesson 337: "My sinlessness protects me from all harm."

Smith says this is the Bible "upside-down."

Deepak Chopra from India is another best-selling author whose primary message is that all men and women have the potential to realize his or her divinity. He says that everyone should recognize that we are gods and goddesses and the most important thing in life is to discover that truth.

Chopra wrote in his book *The Seven Laws of Spiritual Success* that "we will remain unfulfilled unless we nurture the seeds of divinity inside us. In reality, we are divinity in disguise, and the gods and goddesses in embryo that are contained within us seek to be fully materialized."[8]

Chopra says he does not believe in any messiah and I take that to mean Jesus.

"One Jesus is historical and we know next to nothing about him," Chopra says.

He even asserts that the historical Jesus was a follower of the New Age philosophy.[9]

"The salvation Jesus offered was the same as Buddha's: release from suffering and a path to spiritual freedom, joy and closeness to God," Chopra said.

He does not believe that Jesus was the incarnate Son of God who came to earth to die on the cross to pay the penalty for man's sins.

"Christianity is burdened by some impossible expectations," he says, "and one of them is that God forgave all sins through Jesus."

He asserts that Jesus discovered his inner divinity through his own efforts and all of us can do the same.

"In order to advance the New Age version of Jesus…

Chopra asserts that the orthodox version of Christianity is nothing more than a sham perpetrated through the centuries by a self-seeking church with its own suspect agenda," said Kelley Boggs writing for Baptist Press. "In essence, the Christ that is worshipped today by millions world-wide is a false version of Jesus…"

Boggs raised the question: What if an evangelical Christian said that about another religion? "He or she would be branded as a narrow-minded bigot and figuratively stoned in the mainstream media."

Columnist Penna Dexter says the Oprah Winfrey cult and the self-centered New Age religion she supports is dangerous.

"Oprah knows we are spiritual beings – and she's got the money and the stage upon which to elevate her hand-picked spiritual guides and gurus who pull together their versions of truth from various religions and philosophies," Dexter says. "Her beliefs are a far cry from biblical Christianity."

So Oprah's *gurus* have spoken – we are one, we are divine, we are gods. There is no personal God and Jesus came to earth only to show man his built-in deity. The concept of reincarnation gives mankind all eternity to achieve perfection.

Rev. Jeremiah Wright, Jr., Oprah's former pastor when she attended the Trinity United Church of Christ in Chicago, made the following observation concerning her religious transformation: "She has broken with the [traditional faith]… She now has a sort of 'God is everywhere, God is in me, I don't need to go to church, I don't need to be a part of the body of believers, I can meditate, I can do positive thinking' spirituality. It's a strange gospel. It has nothing to do with the church Jesus Christ founded."[10]

Eckhart Tolle, author of the bestselling book *A New*

Earth, has also had a profound influence on Oprah. Together they conducted a live seminar on the Worldwide Web where they discussed each chapter of the book.

The major theme of Tolle's teaching is human divinity and oneness with God.

According to Steve Cable of Probe Ministries, large numbers of Christians are reading Tolle's book and tuning in to the on-line seminar. And he asks the question: Why?

"First, the pervasive influence of post-modern tolerance continues to undermine commitment to the truth of the gospel even in evangelical circles," Cable says. "We are constantly assailed with the message that it is hateful and intolerant to believe that Christianity is true and other religions fall short."

According to Cable, Tolle's *A New Earth* offers a "greater personal peace and an escape from unhappiness" so it's conceivable that many Christians would want to find a way to add the teachings to their Christian doctrines.[11]

"Second, *A New Earth* contains nuggets of truth about the nature of the body, soul and spirit and some practical ideas which may often prove helpful in dealing with anxiety, anger and other issues people face," Cable says. "Tolle is correct in pointing out that our individual and collective selfish egos introduce a lot of pain and suffering into this world."

Also, some Christians accept Tolle's work because he uses quotes from Jesus and others in his book in an attempt to convince the reader his worldview is consistent with Christianity, according to Cable.

However, Tolle's teachings are often diametrically opposed to the scriptures.

For instance, he calls God an "impersonal life force."

I must confess I have no idea what he means by this terminology. He used three words: "impersonal," "life," and "force." That could mean anything and, as I study Tolle, I am convinced he has his head way up in the clouds.

"The view of God as an impersonal life force living in all things is directly counter to the Biblical revelation of God," Cable says. "According to the Bible, God is the creator of the universe not a part of the universe."

In his book that has sold millions, Tolle distorts the concept of sin and evil. He believes that it is impossible to distinguish good and evil since they both originate from the same life force, Cable said.

Concerning salvation, Tolle believes men and women cannot be separated from God for they are a part of God.

"You do not become good by trying to be good, but by finding the goodness that is already within you, and allowing the goodness to emerge," he says.

Then he makes a most astounding statement: "We become a new alive person, not through faith in the atoning death and empowering resurrection of Jesus, but rather through a process of becoming aware of our real self which has been masked by our ego."

Tolle points out that when a person dies he no longer exists as an individual but returns to the universal life force. "So, regardless of what we do or believe during our earthly existence we all have the same ultimate destiny."

Concerning Jesus Christ and Christianity, Tolle believes Jesus certainly was enlightened but he was no more God than any other human "but he was aware that he was a part of the One Life Force which He identified as God."

Cable points out that *A New Earth* is really nothing new but a rebirth of Eastern mysticism which denies the existence of a transcendent Creator God, individual accountability for sin, the existence of the eternal soul, and the need for a Savior. Tolle's spiritual teaching "will result in eternal separation from God for non-Christians and fruitlessness for Christians taken captive by this unbiblical worldview."

Oprah said one of her favorite quotes from Tolle's book is in chapter one when he wrote: "Man made god in his own image, the eternal, the infinite, and unnamable was reduced to a mental idol that you had to believe in and worship as my god or our god."

Oprah and Marianne Williamson have launched the most ambitions New Age project in history on the Oprah and Friends Radio Network on XM Radio Satellite.

The program – *A Course in Miracles* – began on January 1, 2008, and the year-long course was scheduled to be broadcast every day.

"What is a miracle? It is a shift in perception from fear to love," says the promo for the daily program. "And that shift will change your entire life."

According to the promo, the course offers the beginning of a new spiritual journey for everyone and will lead to happiness and peace.

Exactly what is *A Course in Miracles?*

Helen Schucman, Ph.D, received the Course through "inner dictation" during a period from 1965 to 1972. Schucman was a clinical research psychologist at the Columbia-Presbyterian Medical Center in New York City.

She said she experienced the process as one of a distinct and clear dictation from an inner voice, which earlier had identified itself to her as Jesus.

Taking dictation from the inner voice triggered several mystical experiences in her life including dreams, psychic episodes, visions and an inner voice, the Foundation for Inner Peace reported.

Later she shared the experiences with Dr. William Thetford, Ph.D, a professor of Medical Psychology at the Presbyterian Hospital in New York City. After hearing about the dictation from Schuchman's inner voice, he agreed to transcribe the dictation by typing the material taken from her notes.

"The experiences also became increasingly religious, with the figure of Jesus appearing more and more frequently to her in both visual and auditory expressions," the Foundation said.

During the evening of October 21, 1965, the now-familiar inner voice said to her, "This is a course in miracles, please take notes."

"The Voice made no sound, but seemed to be giving me a kind of rapid, inner dictation which I took down in a shorthand notebook..." she said. "It could be interrupted at any time and later picked up again. It made obvious use of my educational background, interests and experience... Certainly the subject matter itself was the last thing I would have expected to write about."

Schucman said taking the dictation from the inner voice made her quite uncomfortable but she was determined to continue.

"It seemed to be a special assignment..." she said.

Thetford once remarked that transcribing *A Course in Miracles* changed his life.

"I recall typing the first fifty principles on miracles... and realized that if this material was true then absolutely everything I believed would have to be challenged – that I would have to reconstruct my whole belief system," Thetford said.

The Foundation for Inner Peace published *A Course in Miracles* in 1972 and thousands purchased the book. It now has been translated into 17 foreign languages.

Even a cursory look at the *Course* reveals that it is an attempt to reinvent and redefine the Person and the work of Jesus through mind training and thought reversal.

All the words of the so-called "Inner Jesus" who allegedly dictated the *Course* to Professor Schuchman and declared there is no sin, no evil, no devil, is quite different from the incarnate Son of God.

However, Oprah said she believed the principles Williamson shared with her from the *Course* could change the world and she still believes it today.[12]

A few years ago Oprah was the catalyst for another spiritual phenomenon known as The Secret. A lady by the name of Rhonda Byrne claimed she had found an important secret: That a person could think about something and it would come true.[13]

"A Porsche, a cancer-free body, whatever," Penna Dexter wrote. "The Secret aired on Oprah's program and was lapped up by consumers."

Then came the book and, with Oprah's endorsement, it became an overnight success Secret groups were formed all over the country.

"Millions of Americans are intrigued with this idea that our thoughts create things," Dexter said. "It's another narcissistic, self-centered lie that denies the sovereign, all-powerful creator God."

Dexter acknowledges that women all across America love Oprah. "Fine, try her diet tips. Emulate her generosity but remember, everything about Oprah's message is self-centered…"

Dr. Albert Mohler wrote a commentary titled *The Church of Oprah Winfrey – A New American Religion?* Dr. Mohler is the president of the Southern Baptist Theological Seminary in Louisville, Kentucky, and a renowned writer and theologian.

He asserts that Oprah has effectively substituted psychology for theology and, hence, has become the high priestess of American society.[14]

"When she features prominent New Age figures on her television show, she helps to mainstream New Age influences and philosophies among millions of Americans," he said. "Her substitution of spirituality for biblical Christianity, her promotion of forgiveness without atonement, and her references to a god 'without labels' puts her at the epicenter of a seismic cultural earthquake."

But Mohler says that Oprah cannot be ignored.

"Oprah's newly-packaged… spirituality is tailor-made for the empty souls of our postmodern age," Mohler said.

"She promises meaning without truth, acceptance without judgment, and fulfillment without self-denial."

Oprah is a charismatic woman who has done many good works with her great wealth. Hopefully, one day she will recognize the hidden dangers of the New Age religion she has so willingly embraced.

Chapter Fifteen
The War Against God

Through the years, most people in this land have expressed an awesome respect for God. That respect was rooted in the nation's religious culture. Even unbelievers and professed atheists chose their words carefully when the public discourse turned to "In God We Trust," "One Nation Under God" or the affirmation in our Declaration of Independence that we are endowed by our "Creator."

However, sadly we have seen the rise and fall of the respect for Deity that has led to a concerted campaign to erase all references to God from the public arena and to the rise of in-your-face atheism.

Almost every day the atheists in this land get offended by something: A young person praying during a high school graduation program, prayers at football games or other sports events, the Christian Cross or the Jewish Star of David, Christmas carols, menorahs, the Ten Commandments or the "one nation under God," in the Pledge of Allegiance. The atheists also are offended if a high school science teachers mentions the possibility of a creator thus questioning evolution, or if a soldier says "God bless you" at a funeral for one of his buddies, or at the Boy Scouts who promise to "Do my duty to God and my country," and finally by a cross on a Veterans' memorial.[1]

Bestselling author William J. Federer points out that these atheists, when offended, "intimidate, threaten and sue to have their will enforced as law."

This is a remarkable development since only 1.6 percent

of the people in this land are actually "card carrying" atheists.

"Imagine that – 1.6 percent atheists, yet their beliefs are becoming the law of the land," Federer said. "A minority forcing its will on the majority – isn't this the classic definition of tyranny?"

President Franklin Delano Roosevelt once said that the early Americans fought a Revolution so the people could be their own rulers.[2]

Thomas Jefferson said that every provision of government is based on the will of the people.

I take that to mean the will of the majority of the people.

"Are we still our own rulers...?" Federer asks. "If the will of majority of 'the people' is not reflected in the laws, then the country is no longer 'democratic,' but has devolved into a tyranny, where a minority forces its will upon the majority."[3]

On April 15, 2010, Federal Judge Barbara Crabb of Wisconsin ruled that the law that sets aside a national day of prayer is unconstitutional for it violates the First Amendment's prohibition of government promoting religion.

The judge ruled on the law after a group known as the Freedom From Religion Foundation made up of atheists and agnostics challenged the law in 2008. They used the worn-thin argument that the law violates the concept of the separation of church and state.

Ronald Reagan once said, the Constitution was never intended to keep people from praying, but to protect their right to pray.[4]

"The frustrating thing is that those who are attacking religion claim they are doing it in the name of tolerance and freedom and open-mindedness," Reagan said. "Question: Isn't the real truth that they are intolerant of religion?"[5]

According to Federer, Reagan also said that the refusal to allow religious exercises such as praying in any area of American life is not "the realization of state neutrality, but rather as the establishment of a religion of secularism."

Federer said that when we examine "ACLU lawsuits, hate-crime legislation, secular revisionism and activist judges trumping the will of the people – one wonders about the future of Lincoln's hope: That government of the people, by the people, for the people, shall not perish from the earth."

Here are some examples of the war against God.

American Idol is one of this country's most-watched television programs. In early April of 2008 in the grand finale for *American Idol Gives Back,* eight *Idol* finalists sang a song entitled "Shout to the Lord" written by Darlene Zschech of Australia's Hillsong Church.

The lyrics open with "My Jesus, My Savior, Lord there is none like you…"

However, the *Idol* contestants replaced "My Jesus" with "My shepherd."

Why are people so afraid of the name of Jesus?

The Associated Press reported that several years ago the Spokane, Washington, public schools provided a calendar of events for the parents of elementary school students for the month. The calendar highlighted Hanukkah, Human Rights Day, Kwanzaa and the Islamic holy day Eid-al-Adha. But it did not mention Christmas.

Walt Disney was a beloved American icon who brought joyful characters to the big screen, characters we loved like Mickey and Minnie Mouse, Donald Duck and Goofy. And Snow White and the Seven Dwarfs, Bambi and Dumbo.

When Promenade Pictures produced and released the animated movie *The Ten Commandments,* the company purchased advertising on Radio Disney, a subsidiary of the Walt Disney Company.

But an ad exec from Radio Disney informed the company that the phrase that Moses was "chosen by God" in the movie could not be broadcast on any of the Disney stations for, I presume, they thought the reference to God might offend some of their listeners.

There's a beautiful song entitled "In God We Still Trust" performed at a Diamond Rio concert. The crowd loved the song and gave the group a standing ovation. However, some radio stations refuse to play the song they believe to be politically incorrect with the reference to God.

Diamond Rio continues to sing the song and it's refreshing to find a group that refuses to pander to the politically correct radio stations that want to erase all references to God from the music world.

Here are the words to the song:

In God, we still trust,
Here in America,
He's the one we turn to every time the going gets rough,
He is the source of all our strength,
The one who watches over us
Here in America, in God, we still trust.

So, what do you think?

Jay Leno, the host of the *Tonight Show*, made a very profound statement regarding the Pledge of Allegiance.

Leno, who at times can be quite irreverent, at least understands that something is eating at the spirit of the people of this land.

He said: "With hurricanes, tornados, fires out of control, mud slides, flooding, severe thunderstorms tearing up the country from one end to the another, and with the threat of bird flu and terrorist attacks, are we sure this is a good time to take God out of the Pledge of Allegiance?"

The famous singer Pat Boone, now a columnist, says there is an "insidious virus of apathetic Americanism, a shedding of traditional mores and beliefs" that has combined to bring our society to the brink of apocalypse.

Ben Stein, who is Jewish, made the following statement

during a *CBS Sunday Morning* commentary: "I don't like getting pushed around for being a Jew, and I don't think Christians like getting pushed around for being Christians. I think people who believe in God are sick and tired of getting pushed around, period."

"I have no idea where the concept came from that America is an explicitly atheist country. I can't find it in the Constitution, and I don't like it being shoved down my throat."

Anne Graham Lotz, the daughter of evangelist Billy Graham, during an interview with Jane Clayson on the *CBS Early Show,* was asked why, if God is good, would he allow a great tragedy such as Hurricane Katrina?

Ms. Lotz reflected on the question, then replied, "I would say... for several years now Americans in a sense have shaken their fist at God and said, 'God, we want you out of our schools, our government, our business, we want you out of our marketplace.' And God, who is a gentleman, has just quietly backed out of our national and political life, our public life. Removing his hand of blessing and protection. We need to turn to God first of all and say, 'God, we're sorry we have treated you this way and we invite you now to come into our national life. We put our trust in you.'"

Meanwhile, four U. S. generals and three other officers are in hot water for promoting a an evangelical organization known as Christian Embassy founded by Bill Bright of Campus Crusade for Christ.

CNN reported that the inspector general determined that the seven officers engaged in unauthorized conduct when they appeared in a promotional video for Embassy.[6]

However, Lieutenant Colonel (Retired) Bob Maginnis believes the investigation is a witch hunt.

"This is an ACLU-type hit on the Pentagon insisting that the inspector general conduct an investigation... whether or not the generals improperly used their influence with regard to a particular ministry," McGinnis said.

According to McGinnis, a high ranking official in the Department of Defense attended a ceremony at the Quantico Marine Corps Base for the dedication of an Islamic Prayer Center. But apparently the inspector general approved that ceremony.

"As far as I'm concerned, he was giving tacit endorsement of Islam just by his presence and making a formal visit," McGinnis said.

We're also witnessing a new phenomenon in the Twenty-First Century: the rise and popularity of angry, in-your-face atheists who have written best-selling books.

"This is atheism's moment," said David Steinberger, chief executive officer of Perseus books, commenting on the success of atheistic bestsellers like *God is Not Great: Why Religion Poisons Everything,* by Christopher Hitchens and *The God Delusion* by Oxford evolutionary biologist Richard Dawkins.

Both Hitchens and Dawkins are from England but have a large following in the United States.

Hitchens' newest book is entitled *The Pocket Atheist* and presents the writings of several famous atheists from throughout history.

"In earlier eras, atheists were on the fringes of society, mistrusted by the mainstream," *Whistleblower Magazine* points out. "Those few who dared to publicly push their beliefs on society, like Madalyn Murray O'Hair, were widely regarded as malevolent kooks."

However, Kitchens' book was No. 1 on the *New York Times'* bestseller list and was at the top of the nonfiction charts for several months. The book condemns Christianity as "violent, irrational, intolerant, allied to racism, tribalism, and bigotry, invested in ignorance and hostile to free inquiry, contemptuous of women and coercive toward children."

Rather than rational, tolerant inquiry and debate, this new breed of atheist writers is filled with hatred toward God and determined to destroy the church and Christians along the way.

"Indeed, arrogant denial of God and condemnation of religious people characterize today's popular atheist books," *Whistleblower* says.

Movie critic Michael Medved says that Hitchens has been "abundantly blessed" for his attack on God. "His outrageous and entertaining book... has become a major bestseller and earned its sardonic author more than a million dollars, according to a recent estimate by the *Wall Street Journal*."[7]

Other bestsellers include *Letter to a Christian Nation* by Sam Harris, sequel to *The End of Faith*, his earlier book; *God: the Failed Hypothesis, How Science Shows that God Does not Exist*, by Victor J. Stenger; *Breaking the Spell: Religion as a Natural Phenomenon* by Daniel C. Dennett; and *Atheist Universe: The Thinking Person's Answer to Christian Fundamentalism* by David Mills.

Apparently many Americans are intrigued with atheism. Dawkins is even selling young people "Scarlet Letter" tee-shirts with a giant "A" for atheist on his website and bumper stickers, too.

Atheists also are promoting their "Blasphemy Challenge." *ABC News* says it is a way "to challenge people to make videos of themselves denying, denouncing or blaspheming the Holy Spirit, and then post them on YouTube." The news giant also calls it "the cutting edge of a new and emboldened wave of atheism."

"The 'Blasphemy Challenge' targets teens and the film *The Golden Compass* was produced to appeal to children," *OneNewsNow* says.

Compass, a film by New Line Cinema, is based on a book written by English atheist Philip Pullman.

Pullman, who has declared that he greatly dislikes the

works of C. S. Lewis – *The Lion, The Witch and The Wardrobe* and others – has written *Dark Materials* which are quite opposite to the Lewis books and J. R. R. Tolkien's *The Lord of the Rings.*

"While Lewis and Tolkien wrote stories imbued with Christian imagery, Pullman's trilogy – which has sold millions of copies and won numerous literary awards... – depicts the death of God and the creation of a 'Republic of Heaven' that has no need for a King," writes Peter T. Chattaway in his article "The Chronicles of Atheism."[8]

Both Lewis and Tolkien always wrote within a Christian framework but Pullman's trilogy tells of a former nun explaining to children that she left the Christian faith because it's "a very powerful and convincing mistake, that's all."[9]

"Some Christians have expressed concern that if *The Golden Compass* is successful it will lead to films based on the other two *Dark Materials* Books, *The Subtle Knife* and *The Amber Spyglass* both of which traffic much more explicitly in the death-of-God theme.

"Atheism is chick, it's cool, it's the latest craze," says Fox TV personality Bill O'Reilly. "The bookstores are chock full of authors declaring 'God Is Not Great,' that God is a 'Delusion,' that you are a moron if you believe in the Deity."

The print and broadcast media seem to approve of these books and give them good reviews.

"Some of the books are also selling very nicely, as it's been a long time since atheists had much to cheer about," O'Reilly says.

This trend toward atheism is not just in books as Hollywood is also involved.

"According to the book *Celebrities in Hell,* a number of big stars may be aligned with the universe, but not with the force that some believe created it," O'Reilly said.

The book, by Warren Allen Smith, includes information on atheists, agnostics, humanists and other skeptics.[10]

They include George Clooney, Angelina Jolie, Carrie

Fisher, Woody Allen, Marlon Brando, George Carlin, Marlene Dietrich, Jodie Foster, Katharine Hepburn, Sally Jesse Raphael, Christopher Reeve, Howard Stern, Uma Thurman and a host of others.

The book quotes the following:

- George Clooney: "I don't believe in heaven or hell. I don't know if I believe in God."
- Angelina Jolie: "There doesn't need to be a God for me."
- Carrie Fisher: "I love the idea of God, but it's not stylistically in keeping with the way I function."
- Former Vice President Al Gore (now himself a Hollywood personality) once said: "Refusing to accept the earth as our sacred mother... Christians have become a dangerous threat to the survival of humanity. They are the blight on the environment and to believe in Bible prophecy is unforgivable."

"Believing in God is not very stylish in mainstream media circles these days," O'Reilly says.

During a debate between O'Reilly and Dawkins who wrote *The God Delusion*, the scientist said that science can explain everything.[11]

O'Reilly challenged him by saying the earth had to come from somewhere. "And that is the leap of faith you guys (atheists) make, that it just somehow happened," O'Reilly said.

Dawkins replied: "You're the one who needs a leap of faith, the onus is on you to say why you believe in something... you believe in, presumably, the Christian God Jesus."

"Jesus is a real guy," O'Reilly replied. "I know what he did... I'm throwing in with him rather than throwing in with you guys, because you guys can't tell me how it all got here."

"We're working on it," Dawkins said.

"When you get it," O'Reilly shot back, "maybe I'll listen."

Atheists in the town of Vernon in Connecticut have taken advantage of the town's policy of allowing holiday-season displays in the public park. They erected a ten-foot tall sign, celebrating the winter solstice. The sign carries a message that blames religious believers for the attack on the Twin Towers on 9/11.[12]

A spokesman for the atheists' organization says that the Twin Towers would still be standing were it not for religion, *World Net Daily* reported.

There have always been atheists in America but most of them chose to keep a low profile. However, today's atheists such as Dawkins, Hitchens, and Harris are a new and different breed. Unlike the atheists of yesteryear, the new atheists not only deny the existence of God, they hope to wipe out all religion.[13]

"People of faith," Hitchens says, "are in their different ways planning your and my destruction, and the destruction of all... hard-won human attainments... Religion poisons everything."[14]

"The way Hitchens lumps all religions and all believers into one category here is typical of his tone throughout the book, and typical of anti-theists in general," says Christian apologist Chuck Colson. "They don't argue; they yell. They've decided that, simply because they dislike religion, there is no reason to respect it. In their minds, it's stupid, dangerous, and that's all that needs to be said."

Colson says that although the anti-theist movement is quite popular today with all the bestselling books, he believes it is doomed to fail. "The moment you take it seriously and start to study it, it falls apart," he says. "There's no substance, just anger and a lot of hot air. Because anti-theists simply ignore evidence and arguments they don't like, they're ill-equipped to deal with them rationally."

"Most traditional atheists simply had their own belief system, and if we wanted our belief system that was okay," Colson points out. "The new breed reflects the death of truth. They're like the communists who feared religion more than anything else because it was a competing truth claim. The Star of David and the cross have been scandalous to every totalitarian leader."

Any individual can choose to be an atheist. But I ask a question: Who gave anyone the right to make history conform to the atheistic worldview?

There are groups in England that say the Boy Scout Oath discriminates against atheists by requiring them to swear an oath to God.[15]

During the 100-year history of the Boy Scouts in England, the young men have promised to do their duty to God and the Queen and that promise has been as important to the scouting movement as jamborees and the three-fingered salute.[16]

"Now, however, it has become the latest target of the ... National Secular Society and the British Humanist Association..." said Jonathan Petre, writing for *The Weekly Telegraph* of London. "They are furious that the Scout Association is refusing to scrap the pledge required of every new member, which they said was excluding a growing number of children without belief."

The two secular organizations, in a letter to the Scout Association in England, said the association's pledge to God is "completely unacceptable" for an organization "that is so committed to personal development of young people and that claims to foster mutual understanding between different beliefs, which of course should include those of no belief."[17]

Stephen Peck, the scouting association's director of programs and development, replied, "It is fundamental to scouting that young people are helped to understand their spirituality. It is in our lifeblood."

All of us remember and admire Judge Roy Moore, the courageous Chief Justice of the Alabama Supreme Court. He fought a long but losing battle to keep a monument to the Ten Commandments in the lobby of the Supreme Court building in Montgomery.

When Judge Moore visited Independence Hall in Philadelphia where the Declaration of Independence was adopted on July 4, 1776, he was shocked as he listened to the tour guide's distortion of the history of that era.

"For example, the tour guide said that the ideas of the Enlightenment and Humanism were instrumental in the formation of our government, but he made no mention of God or his providential role in the affairs of the times," Judge Moore said in a column entitled "Our Moral Foundation." "Yet, according to Thomas Jefferson, the author of the Declaration, it was the law of God which gave us a right to exist as a nation and the purpose of government was to secure those rights which God gave us. Twelve years later, James Madison, the recognized 'Father of the Constitution,' stated that it was impossible for men not to see 'a finger of that Almighty hand' in the drafting of our Constitution."

The judge says that perhaps the most blatant example of those who want to revise history can be found in the Supreme Court Building in Washington. "Directly over the head of the chief justice in the courtroom is a classical display of a tableau containing Roman numerals I through X," Judge Moore says. "A... handbook explains the display is a 'tableau of the Ten Commandments.'"

"Obviously, the present... Court is embarrassed to admit that the Ten Commandments hang above the bench while it allows subordinate federal courts across the land to forbid such displays in school and courtrooms of our country," he writes.

Santa Claus has become the politically correct Crown

Prince of Christmas and is worshipped by millions all across this land while Jesus is virtually forgotten and his incarnation into humanity barely understood as Christmastime has become a quest for personal pleasure and possessions.

It makes me want to ask: Whatever became of Christmas?

ABC's Barbara Walters enjoys receiving Christmas cards from friends but criticized President and Mrs. George W. Bush for including a Bible verse on the card they sent her.

"This is what interested me, that it is a religious Christmas card," she told the other members of *The View* television program. "Usually in the past when I have received a Christmas card, it's been 'happy holiday's' and so on."

Then she read the passage of scripture from the Book of Nehemiah which said: "You alone are the Lord. You made the heavens, even the highest heavens, and all their starry host, the earth and all that is on it, the seas and all that is in them. You gave life to everything, and the multitudes of heaven worship you."

"Now does this (card) also go to agnostics and atheists and Muslims...?" She asked.

Walters also showed the co-hosts a Christmas card she received from entertainer Elton John and his boyfriend with angels on the cover but the angels didn't seem to offend her.[18]

Hundreds of major corporations, retail stores, governmental subdivisions and even those in the media have decided that the word "Christmas" has no place in today's culture.

"Over the last several years, a time of year that was traditionally a period of goodwill and universal brotherhood... Christmas has become an annual battleground between decent people and a relatively small number of secular leftists who insist on carrying on as if auditioning for the role of Scrooge," said columnist Burt Prelutsky in his article "Have Yourself a Dreary Little Christmas."

A member of the city council in New York City has challenged a policy that both allows and encourages Muslim, Jewish and secular decorations in the city's schools during the Christmas season, but does not allow Christian symbols.[19]

Apparently the campaign against Christmas has no age barriers.

The manager of the Plant City Living Center in Florida has forbidden an 85-year-old grandmother from displaying any religious symbols associated with Christmas in the common area of her government-subsidized apartment building.[20]

Also, if the elderly residents plan a Christmas party in the community room, they are not allowed to call it a "Christmas party" but must substitute the word "holiday."

Each year a Sunday school class from a nearby church hosts a "Hanging of the Greens" and Christmas party for the residents. The highlight of the party is when someone places the angel on top of the Christmas tree.

However, the government, heir-apparent to Dickens' Ebenezer Scrooge, said there would be no more angels.

Please, someone, say it isn't so!

Where were all the conservative leaders we sent to Washington to stop this kind of foolishness? Where were the president and vice president when this happened? Where were our senators and congressmen? Why are they silent?

I would think that those in government should be trying to solve some of the bigger issues facing our nation such as rising gas prices, hordes of illegal immigrants crossing our borders every day, the burgeoning federal budget, waste in government spending and other problems. Rather, they are spending their time making sure the old folks in a retirement center don't place an angel on a Christmas tree.

Each year the Culture and Media Institute (CMI) issues

the "Grinch-0-Meter with information about those who are trying to secularize, trivialize and abolish Christmas.

The Grinch-0-Meter cited Old Navy's "open your gifts on holiday morning" and the city of Virginia Beach, Virginia's decision to rename its Christmas Festival the "McDonald's Holiday Lights at the Beach Brought to You by Verizon Festival."

Here are three others that made the Grinch-O-Meter, according to CMI:

- "Planned Parenthood – the abortion provider – sent out email Christmas cards to its supporters titled 'Choice on Earth.' The irony of an abortion provider sending out greetings on a day that celebrates the Birth of Christ is astounding," said Kristen Fyfe of CMI.[21]
- "Anyone and everyone, especially Sherri Shepherd and Joy Behar of ABC's *The View*, have complained about Mike Huckabee's Christmas commercial because it mentions 'Christ' and 'Christmas.'"[22]
- "The Catholic League reports that Hanukkah is celebrated at Harvard University, but not Christmas. A giant menorah sits in Harvard Yard but there is not a nativity scene in sight.
- Additionally, Pembroke Lakes Mall in South Florida displays a five-foot-tall menorah but all nativity scenes have been censored."[23]

Chapter Sixteen
The Late, Great
American Church

I don't pray that God is on our side.
I pray that we are on God's side.
~ Abraham Lincoln

It may come as a surprise to many of you that America does not have a favored-nation relationship with God.

But you say, Bill, the Christians in this country send more missionaries into the world to preach the Gospel than any other nation; we give more to charity; and often our men and women form the first-response teams to try to alleviate suffering during natural disasters anywhere in the world. We feed the hungry homeless, minister to those in prison, provide homes for unwed mothers, have programs to help alcoholics and drug addicts, and a hundred other ministries. We also have more Bibles, Christian books, magazines and bookstores, religious radio and television programs and inspirational music than all the other nations of the world combined. And we've invested trillions of dollars in mega-churches and lesser churches where millions worship on Sunday mornings.

However, one of the great mysteries of our time is why all these great churches, wonderful ministries and biblical resources have only a minimal "salt" and "light" influence on our culture. We have more abortion, pornography and

224 of the world

sexual perversion, drug addiction and alcoholism, divorce and gambling than most of the other countries of the world. And we are living in a country where high courts in two states have defined deviancy down and ruled in favor of homosexual marriage.

As we examine Christendom in America today, we see that many of the indicators of a healthy, vibrant church are in decline.

Three major public opinion polls reveal that American teenagers are leaving the church in great numbers, and not coming back. The polls were taken by *USA Today*, the Southern Baptists' Lifeway Christian Resources and Cornell University. They all come to the same conclusion: that the future of Christianity is in crisis.

"For example, when you travel to countries that were once bastions of Christendom, like Holland, Scotland, Switzerland, Germany and Australia, you find that the Christian population has dwindled to an insignificant, impotent minority while the People of the Lie have become the standard bearers of the culture..." said Syndicated Columnist Dr. Ted Baehr. "For instance, now in Germany, there are fewer Protestants than Muslims, and one leader in Berlin commented that it would be better for Germany to be taken over by the Muslims than Christianity to return. Aside from arguing with this ignorant person that everywhere Islam has advanced has turned into a vast wasteland, his cultural self-hate and rejection of the Christian faith of his forefathers... is just an example of a problem that is occurring throughout Western Civilization."

But you will say, "I don't believe this could happen in our land. Look at all the mega-churches, the television ministries and evangelists like Billy Graham."

Yes, there are myriad ministries but every current survey reveals that in spite of all the ministries we're losing our young people in record numbers and they are the future of the church.

"*USA Today* said many of the teenagers claim they are leaving because the church does not call them to a rigorous, disciplined, stimulating faith-filled life but to a movie night and pizza," Dr. Baehr said.

Ron Luce, founder of the highly successful Teen Mania, says that if our culture continues on its dangerous course, within the next two generations only four percent of the people in this land will be Christian. If Luce is correct, and I believe he is, we have about fifty years before we self-destruct, for a generation is twenty-five years. Sadly, this has already happened in several countries around the world.

Author Bruce Shortt says research reveals that "the overwhelming majority of children from evangelical families leave the church within two years after they graduate from high school" and "only nine percent of evangelical teens believe that there is any such thing as absolute truth..." Shortt is a graduate of Stanford University, has a law degree from Harvard and is a Fulbright Scholar. Here are some of the world's influences molding today's generation of young people: Rap and rock music, HBO and MTV, racy books and magazines, violence-laced video games, movies that glorify sex, school textbooks that teach evolution theory as truth, large billboards and television advertising with young women wearing only limited clothing, teachers and TV talk-show hosts who make fun of sexual abstinence, the pervasive internet with every kind of depravity imaginable and television sports events that now are synonymous with drinking beer.

Dr. James Dobson, child psychologist and founder of *Focus on the Family,* said in May of 2009, "If you've been paying attention to today's youth culture, and especially if you're currently raising kids, you are aware that they are growing up in an environment awash with materialism, illicit sexuality, violence, drugs, cutting, piercing, eating disorders, teen suicide and other symptoms of depression and despair."

According to Dr. Dobson, millions of teenagers are "confused and overwhelmed" by the pressures they face.

"To cite just one example, a phenomenon that the media has dubbed 'sexting' has emerged in the past few years, in which children as young as middle-school are taking nude photos or videos of themselves or while engaging in sexual activity," Dr. Dobson said. "Then they send them to their peers via cell phone."

He pointed out that the National Campaign to Prevent Teen and Unplanned Pregnancy warns us that nineteen percent of teen guys and gals have "engaged in this dangerous activity without knowing they may be creating and distributing child pornography!"

"I tremble for our society when I consider what we have permitted to happen to our children," he said.

Yet millions of Christians seem to be oblivious to the depravity that has captured the souls of a younger generation and many pulpits are silent on the great culture-war issues of our time.

Richard Ross, co-founder of the True Love Waits movement, says the churches are "getting lazy" in promoting abstinence as the only safe method for preventing the spread of STDs.

A report from the U. S. Centers for Disease Control made public in March of 2008 revealed that one in four girls between the ages of 14 and 19 have sexually transmitted diseases and that means 3.2 million are so infected. Permissive sex – fueled by Hollywood and the motion picture industry – has become epidemic in America.[1]

"This is obviously an indicator of the huge number of people – in this case girls, but it obviously takes boys as well – that are engaging in sexual activity in very immoral and unprotected ways, unprotected morally and unprotected health-wise," says Dr. Daniel Heimbach, a professor of Christian Ethics at the Southeastern Baptist Theological Seminary in Wake Forest, North Carolina. "This is a disaster in the making… a road toward destruction."

Parenthetically, Baptist Press reported on June 18 of 2008 that Professor Heimbach has withdrawn his membership from the 1,000-member Society of Christian Ethics after the organization adopted new guidelines that affirm homosexuality.

Society of *Christian* Ethics?

Another report reveals that one in five adults in America has genital herpes, a sexually transmitted disease.

George Barna, perhaps the best-known authority in America on patterns of religious life, says in his book *Revolution* there are changes taking place that will impact every believer in America.

"Committed, born-again Christians are exiting the established church in massive numbers," he says in his book.

Why are they leaving? Where are they going? And what does this mean for the future of the church in America? Barna asks.

There are several reasons.

According to a recent survey by LifeWay Research of the Southern Baptist Convention, more than two-thirds of all Protestant young people in America between the ages of 18 and 22 are walking away from the church.[2]

According to *Christian News Today*, some of those return to the faith but others leave permanently.

Another LifeWay Research study reveals that most of the people joining churches today are already church members looking for a new congregation and that it is unlikely that most unbelievers will ever visit a church.[3]

Ed Stetzer, director of LifeWay Research, says it is apparent that our culture is growing increasingly resistant to Christianity.

"Believers must resolve to step into their world to share the Good News…" Stetzer said. "If we are waiting for them (unbelievers) to someday walk into our churches, that someday may never come."

LifeWay also reported that the number of people baptized in Southern Baptist Churches in 2007 was the lowest since 1987 and fell for the third straight year.

"This report is truly disheartening... We are a denomination that, for the most part, has lost its evangelistic passion," said Tom Rainer, president of LifeWay.

Although LifeWay does not break down the numbers, many of those baptized in 2007 were children of church members who experienced salvation through such programs as AWANA that helps churches minister to young people, Vacation Bible School and youth camps.

That means that fewer and fewer adults are embracing the Christian faith and the numbers are decreasing every year as the ranks of those who never attend church are growing.

The Pew Forum on Religion and Public Life interviewed 35,000 adults and learned that one-quarter of them have departed from the faith of their childhood for another religion or no religion at all.

The numbers of those who never go to church are growing three times as fast the number of unbelievers visiting the church, according to the report.[4]

Shouldn't that prediction serve as a wake-up call for every Christian in America?

According to Janice Crouse of Concerned Women of America, thousands of young people are attracted to and experimenting with the Wicca religion. Wicca is a nature-based religion that includes pagan witchcraft and magic.

Crouse wrote in *Religion Journal* that youth pastors in churches throughout the land are concerned that the young people in evangelical churches are also taking part in Wicca.[5]

She points out that Wicca "involves nature worship, stresses moral autonomy, and includes remedies and spells – beliefs that...are distinctly different from orthodox Christianity..."

Moral autonomy or the idea that "nobody can tell me what to do" is very appealing to young people who "don't want the church telling them that there are boundaries, there are things that they can't do," Crouse said.

"Some people think this goes back to... many of the games and television programs that feature witchcraft and magic and fairy tales that have a dimension to it that if you pull out some kind of spell you can make anything happen," she said. "This has really become quite entrenched in many of the young people's (church) groups..."

Author and Christian activist Linda Harvey says that "casual occultism" is spreading rapidly throughout the youth culture in this land.

"What's the significance of the sudden appeal of teen paganism, even witchcraft?" she asks in her book *Not My Child: Contemporary Paganism & New Spirituality*. "Is fortune-telling and spell-casting just a new version of teen rebellion or a more sinister development?"

Harvey says "the fingerprints of evil," that often appear to be innocent, can be found in many youth activities and entertainment.

She says that "compromised churches" have contributed to the increased interest in the occult among young people.

What about men?

Author Tristan Emmanuel, founder of Equipping Christians for the Public-Square, has written a book entitled *Why Men Hate Going to Church.* [6]

Emmanuel points out that 90 percent of the men in America believe in God and five out of six say they are Christians.

"But only two out of six attend church on any given Sunday," he said. "The average man accepts the reality of Jesus Christ, but fails to see any value in going to church."

According to Emmanuel, the solution is quite simple:

"Start encouraging the men in the church to be men – not women in drag… Men need to be men again. They need to take up their responsibility the way God intended…"

I'm sure we could find any number of "whipping boys" to blame for the approaching church train wreck that may be lurking just around the bend. We could blame our culture – Hollywood, the corrupt news media, a secularist academia, a hedonistic society, liberal democrats and do-nothing republicans.

But in my opinion, many of the problems we face today can be traced right back to the pulpits of this land, pulpits that are silent on many of the burning issues of the culture war in which we are engaged.

Some pastors are careful not to offend anyone for fear the offended will not come back to church. Hence, they preach a feel-good religion that borders on fantasy often shaped by a secular progressive worldview that emphasizes success, pleasure and possessions. And apparently that message resonates well with millions in America.

It appears to me that some of our pulpits have strayed far away from the culture-warrior-type preaching of John the Baptist and have embraced the psychology of Dr. Phil.

I personally like Dr. Phil, but it is not his responsibility to teach about sin and repentance, separation from the world and holiness before God. That is reserved for men called of God to preach the Gospel.

Many churches no longer have altar calls to give people the opportunity to repent of sins and seek God's forgiveness. I can remember when great conviction would come down on an entire church congregation during a worship service and people young and old would rush to the altar where they would confess, "I'm tired of being a liar, tired of being a thief, tired of adultery."

Genuine repentance and sorrow for sin has virtually disappeared from church life today. And we have thousands and perhaps millions of practical atheists who believe in

God and even attend church but live their lives as though he does not exist.

I often hear the call for repentance and revival today, but I don't see it.

Where are the prophets like John the Baptist, Peter and Paul, John Wesley and George Whitefield, Charles Finney, D. L. Moody, Billy Sunday and Billy Graham and multiplied thousands of others who cried out against sin?

Have they been replaced by those who preach the feel-good Gospel of Success?

Who will offer the world a way out of the madness of today's culture?

J. Lee Grady, former editor of *Charisma and Christian Life* magazine, says God is shaking everything that can be shaken, including our Christian colleges, mega-churches and prominent ministries.[7]

Grady has called for a national spiritual awakening.

"There were seasons in America's past when sinners became so convicted of their sins that they collapsed under the weight of their guilt," Grady said. "During the days of revivalists George Whitefield and Charles Finney, huge waves of conversions led to widespread transformation of society. Drunks became sober, prison inmates sang hymns, stingy business owners stopped oppressing their workers, atheists surrendered their unbelief and rebellious children returned to faith…Can such a movement happen again? It must, or our country will descend into its darkest hour."

Emmanuel says that if there is revival and reformation in the church, there will be revival and reformation in society.

I wonder how many of us are willing to stand on the unchanging biblical values in a changing world? Or has the church raised the white flag of surrender? Taken the Fifth Amendment? Gone AWOL?

The sin of silence is killing the church! That's why we are losing the culture war. Sometimes I wonder if Christians have even heard about the war – the life and death struggle for the future of Christianity in this land.

There is a great deal of turbulence and discord among Baptists, Methodists, Presbyterians, Episcopalians, the United Church of Christ and various other denominations and, of course, Catholics. That turbulence has emasculated the church, weakened its voice and caused it to stand by helpless as this land descends into a morass of a secular, post-Christian society.

What about the Baptists? Former Presidents Jimmy Carter and Bill Clinton and Vice President Al Gore, Baptists all, want to start another church movement and give it the benign name of "A New Baptist Covenant."

Carter is the leader. He called together thousands of delegates, mostly moderate and liberal Baptists, in Atlanta in February of 2008 in an effort to find unity for social action on racial and theological differences among 30 different Baptist denominations.[8]

That sounds like a pretty good idea. But what was the real purpose for the meeting? .

"Conspicuously absent from the gathering was America's largest Baptist and Protestant denomination, the Southern Baptist Convention..." said James A. Smith, Sr. writing for *Baptist Press*. Smith is the executive editor of the *Florida Baptist Witness* weekly newspaper. "The lack of involvement of the SBC, however, was not surprising, given Carter's long history of SBC bashing, including his very public departure from the SBC in 2000..."

According to Smith, some secular journalists attending the meeting reported that its purpose was to find "the liberal answer to the (conservative) Southern Baptist Convention" as a columnist for the *Wall Street Journal* reported.

The New York Times said, "But for other Baptists and experts on the faith, the central aim of the gathering seems to be to create a theological and political counterweight to the

Southern Baptist Convention, which many of the groups that plan to attend have left."

Organizers of the event rejected requests from two homosexual-friendly groups to have exhibits at the New Covenant meeting.

However, American Baptist author Tony Campolo represented the homosexual cause during a speech at the meeting. He wore a brightly colored religious stole to show his solidarity with homosexual and lesbian Baptists.[9]

Carter called for unity on "man-made issues such as the role of women in the church, homosexuality and abortion."

Most Southern Baptists disagree with the former president on various issues including his belief that Mormons are Christians. They also disagree with Clinton who once said he saw God manifested in a voodoo ceremony in Haiti, and Gore's pantheism found in his book *Earth in the Balance: Ecology and the Human Spirit*.[10]

But the turmoil is not limited to Baptists.

In January of 2008, the Reverend Larry Phillips, a United Church of Christ pastor, blessed the new Planned Parenthood Clinic in Schenectady, New York, on the 35th anniversary of the Supreme Court ruling that legalized abortions.

Reverend Phillips, pastor of Schenectady's Emmanuel Baptist-Friedens United Church of Christ, declared the clinic "sacred and holy... where women's voices and stories are welcomed, valued and affirmed; sacred ground where women are treated with dignity, supported in their role as moral decision-makers... sacred ground where the violent voices of hatred and oppression are quelled."

The reverend's blessing on the clinic that will perform abortions sounded to me like he was experiencing some kind of deep spiritual ecstasy.

He said it would be a place where women would be treated with dignity. But he said nothing about the dignity of the little unborn guys whose life will be terminated in the clinic. He also said that the "sacred ground" would be safe from "violent voices of hatred and oppression." I take that to mean people like me, who oppose Planned Parenthood's killing of the unborn, are the violent ones. But there is no more violent place in the world than a mother's womb when an innocent baby is about to die in one of Planned Parenthood's clinics.

There is a battle raging among Episcopalians over the ordination of a practicing homosexual bishop and the homosexual/lesbian issues that dwarf the turmoil being experienced by any of the other denominations.

Some 600 conservative Anglican Churches in the United States and Canada have taken steps to form a new branch of the worldwide Anglican Communion. They will be known as the Common Cause Council of Bishops and the organization will serve as an umbrella for churches opposed to the ordination of homosexual clergy and same-sex marriages.

"Based in England, the Anglican Communion has been on the verge of schism since the North American branch, known as the Episcopal Church, ordained an openly gay man as bishop despite the worldwide church's position that homosexual practice is unbiblical," John Draper and Adrienne Gaines reported for *Charisma* magazine. The Reverend Gene Robinson, who lives with a homosexual lover, was named bishop of New Hampshire.

The Reverend Alan Hansen, who is the president of the Atlanta-based Acts 29 Ministries, an organization seeking to bring renewal among Episcopalians, said he believes God is judging the church.

"The Episcopal Church is saying what is evil is good..." Hansen said.

Some conservative Episcopalians express little surprise that one Episcopal priest in Seattle says she is both a Muslim and a Christian.[11]

Ann Holmes Redding, who has served as a priest for 20 years, claims she became a Muslim in 2006, but continues to embrace the Christian faith.

Redding says she has never believed in the Episcopalian doctrine of "original sin" and has questioned the divinity of Jesus for years. The *Seattle Times* reported she does not believe the church's doctrine of the Trinity and says it is only "an idea about God" and should not be taken literally. Neither does she believe that God and Jesus are the same.[12]

Meanwhile, Redding began teaching New Testament classes at Seattle University in the fall of 2007.

The bishop of the Episcopal Diocese of Seattle said he accepts her as a priest and a Muslim.

Peter Frank of the conservative Episcopal Diocese of Pittsburgh said the people of America should realize "there are a lot of us in the Episcopal Church that really do stand for mainstream Christianity."

But there are times these gallant Episcopalians pay a heavy price for standing by their biblical convictions.

"Just try remaining faithful to God's Word today in the Episcopal Church USA, for example," writes David Kyle Foster, director of Mastering Life Ministries in Nashville. "If you dare try standing with our Lord against the enthronement of unrepentant, practicing homosexuals to the episcopate, you will likely be sued for you church properties, stripped of your salary and pension, and perhaps even defrocked from the priesthood altogether. Such draconian sanctions have already been imposed on faithful priests in that denomination by the very people who cry, 'Peace and unity!' and who 'celebrate diversity and tolerance.' And the purging has only begun."[13]

Foster explains it is okay for Episcopalian priests to deny the deity of Christ or his resurrection or the atonement for sin.

"But dare oppose the enshrining of the immorality that nailed Christ to the cross and there will be hell to pay,"

Foster said.

In his article entitled *The Judas Church,* Foster said the Evangelical Lutheran Church of America is also on the verge of sanctioning homosexual behavior thus joining the United Church of Christ and the Episcopalians.

He laments the fact that many charismatic and evangelical churches are racing toward apostasy.

"Hardly a week goes by before another story of immorality, power grabs or blatant thievery is uncovered among the leadership of one church or another," he said.

How do we Christians respond to the spiritual confusion all around us?

"We separate ourselves from it," Foster said. "We refrain from attending it and supporting it. We stop turning our heads the other way and pretending as if nothing is wrong. We speak up, expose the darkness and declare the truth of God's Word."

Some religious leaders see the same-sex phenomenon as a sign of apocalypse.

Conservatives within the United Methodist Church believe liberals in the denomination are also working to promote the transgender lifestyle.[14]

One issue that concerns the conservatives is the reappointment of a transsexual to be the pastor of a Baltimore Methodist church. The Reverend Ann Gordon has been the minister of St. John's United Methodist Church for five years before undergoing sex-change surgery. After adopting a new name, Drew Phoenix, Gordon was reappointed to St John's in early 2007 by Bishop John Schol.

Mamoud Ahmadinejad, the president of Iran, received a warm reception from the so-called "religious left" during his visit to New York City in 2008.

While in New York, the Iranian president met with 150 church officials at the United Methodist Women's Church Center for the United Nations, according to Mark Tooley, director of the Institute on Religion and Democracy, a renewal group within the Methodist Church.

According to Tooley, several other groups joined the Methodists for the meeting including the Mennonite Central Committee, Jim Wallis' Sojourners, *Pax Christi*, the World Council of Churches, the Church of the Brethren, and the American Friends Service Committee (a Quaker organization).[15]

"There is a blossoming friendship going on between American religious and church officials and Iran's apocalyptic radical president," Tooley wrote in the *Weekly Standard*. "The church officials say they want to avert any major conflict between Iran and the U.S., and they profess to be peacemakers. But in this process of ostensible peacemaking, they seem to want to minimize the Iranian president's very ugly threat against Israel and the U.S."

Tooley said Ahmadinejad must have come away from the meeting realizing he has "reliable American friends who will oppose any strong policies aimed at his regime, while expressing limited concern about his intemperate plans" toward the U. S. and Israel.

"It's very odd how the religious left in America often fawns toward or apologizes for the transgressions of radical Islam, given that radical Islam of course is very hostile to much of what the religious left stands for in terms of women's rights and gay rights," Tooley said. "But the continuum seems to be that the religious left sympathizes with any movement or foreign governments that are hostile to the U.S."

A spokesman for the National Council of Churches (NCC) called the Iranian leader a very "pious" and "religious man."

Are they talking about the Ahmadinejad that I know? The one who denies the holocaust and wants to destroy Israel? And shouts "Death to America?"

What planet is the NCC living on?

In October of 2007, J. Daryl Byler of the NCC wrote a letter to President George W. Bush praising the Iranian leader and calling for a meeting between the two leaders. He said such a meeting could signal "a positive change" in the relationship between the United States and Iran.[16]

However, President Bush chose to honor America's resolve never to meet and negotiate with terrorists.

Pope Benedict XVI's visit to the United States in May of 2008 was a magnificent event with all the accompanying Papal royalty and a smiling vicar who reminds me of every young boy's grandfather.

We Protestants should be thankful for our Catholic brothers who have taken such a strong stand for the lives of the unborn and we praise the Lord for their courage even in the midst of bitter opposition from the secular progressives and the corrupt news media.

I was sickened when liberal talk show host Bill Maher called the good and godly Pope a "Nazi." Mayer for some reason unknown to me has a pathological hatred for all Catholics and Protestants. For instance, after the Reverend Jerry Falwell passed away, Maher said on national television he was glad he was dead.

An HBO executive informed Catholic League President Bill Donohue that Maher was planning to apologize for his "Nazi" comment about the pope. HBO sponsors Mayer's *Real Time with Bill Mayer* program.

Donohue said Maher "lied when he said the pope 'used to be a Nazi.' Like all young men in Germany at the time, he was conscripted into a German Youth organization (from which he fled as soon as he could.)"

"Every responsible Jewish leader has acknowledged this reality and has never sought to brand the pope a Nazi," Donohue said. "That job falls to Maher."

Will the Catholics accept Maher's apology?

"Assuming it comes across as genuine, the answer is

yes," Donohue said. "But I hasten to add that what we would really like to see is for Maher to stop with his hateful diatribes against the Catholic Church."[17]

However, the Catholics have far more important issues to deal with that should eclipse dealing with an angry nobody like Bill Maher.

For instance, I have serious questions about the resurgence of papal authority that is a return to hard-core exclusivism following the more moderate approach of Pope John Paul II. It troubles me that Pope Benedict has "reasserted the universal primacy of the Roman Catholic Church, approving a document that says Orthodox churches were defective and that other Christian denominations were not true churches... and Catholicism provides the only true path to salvation."[18]

Reuters News Service reported that "the Vatican said Christian denominations outside Roman Catholicism were not full churches of Christ."

It seems to me the pope and Catholics everywhere have far more important matters to deal with than whether a Baptist or a Methodist is on the dangerous road to perdition. Although we Protestants have our own disappointments, and there are many, there has never been anything in history to equal the moral failures of tens of thousands of errant Catholic priests throughout the world. Thankfully, this pope is trying to deal with these problems and make restitution to the victims.

Back in 1998, Dr. Phil Fernandes presented a scholarly paper to the Northwest regional meeting of the Evangelical Theological Society titled *Approaching the 21st Century: The Death of God, Truth, Morality, and Man.* Dr. Fernandes is president of the Institute of Biblical Defense and presented the paper at the Multnomah Bible College in Portland, Oregon.

In his paper, he discussed the prophetic insights of Ger-

man atheist Friedrich Nietzsche and two well-known
Christian thinkers, C. S. Lewis and Francis Schaeffer. The
subject? The future of Western Civilization.

Fernandes discussed the nineteenth century's death of
God movement so declared by the atheist Nietzsche. That
was a forerunner of the twentieth century's death of universal truth and absolute moral values which, he believes,
will lead to the death of man in this century unless Christians find some way to reverse the trend.

"Fredrich Nietzsche (1844-1900) proclaimed that 'God is
dead'," Fernandes said. "By this he meant that the Christian worldview was no longer the dominant influence on
the thought of Western culture. Nietzsche reasoned that
mankind had once created God through wishful thinking,
but the nineteenth century man intellectually matured to
the point where he rejected God's existence."[19]

According to Fernandes, Nietzsche reasoned that if the
God of the Bible does not exist, the Judeo-Christian values
taught in the Bible should have no influence on mankind.
Hence, he believed man could determine his own meaning
of life, truth and morality.

"The history of the twentieth century has proven Nietzsche's basic thesis correct," Fernandes said. "Western culture's abandonment of the Christian worldview has led to
a denial of both universal truth and absolute moral
values... The death of God is not a step forward for man; it
is a step backward – a dangerous step backward. If God is
dead, then man is dead as well."

Fernandes points out that the nineteenth century
brought the death of God to Western culture and the twentieth century the death of truth and morality. And, two
great Christian thinkers – C. S. Lewis (1898-1963) and Francis Schaeffer (1912-1984) – believed that the death of man
would follow unless man repents.

Lewis, who wrote the prophetic book *The Abolition of
Man*, predicted that moral relativism would encourage the

rejection of objective truth and would be the downfall of any society that embraced that relativism. Hence "the abolition of man."

"According to Lewis, the rejection of traditional values and objective truth will lead to… the arbitrary wills of the few who rule over the billions…" Fernandes explained.

We are seeing evidence of the C. S. Lewis prophecy all around us.

Fernandes reminds us in his paper that it was Francis Schaeffer who proclaimed that Western culture is now in the "post-Christian era."

"By this he meant the same thing Nietzsche meant when he declared 'God is dead,'" Fernandes said. "Schaeffer was saying that the Christian worldview was no longer the dominant presupposition of Western culture. Now, a secular humanistic view of reality permeates the thought of the West."

Because of that paradigm shift in Western culture, mankind has fallen down into a trap Schaeffer describes as "the line of despair."

"Schaeffer meant that, by throwing the God of the Bible out of the equation, modern man, alone and without divine revelation, could not find absolute truth and eventually gave up his search for it," Fernandes said. "According to Schaeffer, modern man … believes there are no absolutes… has rejected universal moral laws and has embraced moral relativism."

In his writings, Schaeffer noted that when society functions with no fixed ethics, a small minority often decides what is best for the majority and it becomes law.

"Schaeffer compares the present climate of arbitrary lawmaking to the fall of the Roman Empire," Fernandes said. "The finite gods of Rome were not sufficient to give a base in law for moral absolutes; therefore, the Roman laws were lax and promoted self-interest rather than social harmony. This eventually led to a state of social anarchy as violence and promiscuity spread throughout the empire."

Schaeffer saw the 1973 Supreme Court decision to legalize abortion as an example of a minority of judges forcing their will or worldview on the majority in the land.

"But, according to Schaeffer, this is just the beginning, for once human life has been devalued at one stage, then no human life is safe," Frenandes said. "Abortion will lead to infanticide and euthanasia… Schaeffer documents the erosion of respect for human life in the statements of Nobel Prize winners Watson and Crick. These two scientists, after winning the Nobel Prize for cracking the genetic code, publicly recommended that we should terminate the lives of infants, three days old and younger, if they do not meet our expectations."

Schaeffer urged the evangelicals in this land to sound the alarm and called on church and society to repent for, he said, the death of man is fast approaching.

"Learning from the mistakes of the past, let us raise a testimony that may still turn both the churches and society around – for the salvation of souls, the building of God's people, and at least the slowing down of the slide toward a totally humanistic society and an authoritarian suppressive state," Schaeffer said.[20]

Chapter Seventeen
The Philistine Syndrome

The secular progressive establishment in this land is quite formidable. It includes the liberal/progressive members of Congress; left-leaning, socialist academia and news media; the evolutionists and environmentalists; homosexuals/lesbians; global warming alarmists; Hollywood producers; abortionists; feminists; atheists; proponents of amnesty for all illegal immigrants; and powerful new age leaders such as Oprah Winfrey.

When I look at their great political, economic and social power, I feel like little David facing the Philistine giant Goliath.

But regardless of the almost omnipotent power of the secular progressives, we who still believe in biblical values must draw a line in the sand and stand and fight for what we believe. We must get out of our comfort zones and shout to Christians and conservatives everywhere: "Wake up! Stand up! Speak up!"

Some 26-million evangelical Christians failed to vote in the 2006 national election when the people turned over the keys of power to the secular progressives in Congress. And we are seeing the tragic consequences of that election every day.

We must be silent no more.

After an avalanche of bad news, there may be a light at the end of the tunnel.

A group of 150 Christian leaders – Catholics, Eastern orthodox and Protestants – on Nov. 20, 2009, announced the "Manhattan Declaration" dealing with several cultural issues facing our land. They declared a strong commitment to defend the sanctity of human life, biblical marriage and religious liberty without compromise. The announcement was made during a news conference in Manhattan in New York City.

The Declaration also called for civil disobedience in matters of Christian conscience.

The signers of the 4,700-word declaration pledged to work together to "embrace our obligation" to speak out on behalf of the dignity of all human beings, marriage as the union of a man and a woman, and the freedom to express religious convictions.

"We will not comply with any edict that purports to compel our institutions to participate in abortions, embryo-destructive research, assisted suicide and euthanasia, or any other anti-life act," the statement says, "nor will we bend to any rule purporting to force us to bless immoral sexual partnerships, treat them as marriages or the equivalent, or refrain from proclaiming the truth, as we know it, about morality and immorality and marriage and the family. We will fully and ungrudgingly render to Caesar what is Caesar's. But under no circumstances will we render to Caesar what is God's."

Some of the signers include Robert George, a Catholic professor at Princeton; Richard Land of the Ethics and Religious Liberty Commission of the Southern Baptist Convention; James Dobson, founder of Focus on the Family; R. Albert Mohler, Jr., president of the Southern Baptist Theological Seminary; Charles "Chuck" Colson of the Prison Fellowship Ministries; several Roman Catholic bishops; and others.

The Tea Party movement – spreading like wildfire across America – is another indication that the people of this land are waking up and are willing to take a stand against the government's drift toward socialism.

I've heard a lot of criticism of the members of the Tea Party – that they are extremists – particularly from the national news media and leaders of Congress. One political bigwig in Washington had the nerve to tell these people to "shut up!" I thought that quite strange for I have always believed that we Americans have had the right to voice our opinions, particularly when it comes to government.

Senate Majority Harry Reid of Nevada and former Speaker of the House Nancy Pelosi – along with *The New York Times* and NBC News – have insinuated the Tea Party crowd is made up of kooks and terrorists.

Last year I attended a Tea Party rally in my hometown of Longview, Texas, and wish all the politicians so critical of the movement could have been there. It was sponsored by We the People, a national movement that now includes 25 percent of the people of the United States. What did I see? A large crowd. Patriots – many elderly – veterans, businessmen, pastors of churches, the general manager of a manufacturing company, a well-known congressman, young and old, black, white and Hispanic and hosts of others. It was quite clear those people believe in small government, a strong national defense, less spending, free-market solutions to economic growth and that the government should stay out of the healthcare business, the automobile business, the banking business and everything else where they have no business.

What did they all have in common? They were convinced that elected officials in Washington have turned their backs on them, no longer listen to the voices of the people and, hence, are not their friends.

I'm glad I attended the Tea Party rally. It was encouraging to see the American people taking a stand on what they believe. After all, they are paying the bills.

On April 15, 2010, there were 800 Tea Party rallies across America and the movement is growing stronger every day for those rallies are an expression of the people who want to exercise their free speech guaranteed by the Constitution.

I agree with the Tea Party people that freedom must be defended from a government that wants to take it away from us.

An *NBC/Wall Street Journal* poll released on Dec. 17, 2009, revealed that 41 percent of likely voters in this land have a favorable opinion of the Tea Party. That compares to 35 percent who are favorable to Democrats and 28 percent Republicans.

I recently read an open letter to *WorldNetDaily* readers from Managing Editor David Kupelian.

Kupelian asked some very probing questions: "Do you feel as though there's no hope for America? Is the seduction so advanced, the corruption so widespread and the confusion so universal that there's just no way back? Is America the beautiful – (Ronald) Reagan's 'shining city on a hill' – gone forever?"[1]

"And yet, appearances can be deceiving," Kupelian said. "Moreover, if evil can make things appear to be so black that we abandon hope, then we're in real danger of losing a fight we could have won."

He recognizes that today the people of America are facing critical problems. They include the following:

- Fascist Islam determined to radically change America through "terrorism" and "subversion."[2]
- The invasion from the south by millions of illegals who are "changing our country's demographics, values and character" and could change the United States forever.
- "A raging culture war in which perversion and immorality are glorified, while traditional Judeo-Christian values – the stuff of the WWII 'Greatest Generation' – are contemptuously denounced as old-fash-

ioned, ignorant and bigoted."
* Presently, conservatives in the land "are danger-
ously divided and demoralized..."

"Scary, huh?" Kupelian asks. "So how do we deal with
all of this? Do we sit back and wait until tomorrow's paper
comes out and then read what happened? Or do we actually do something?"

There is a solution to the poison that has infected our
culture.

"Although most cultures die from the collapse of character, some cultures are revived," said Baehr in his syndicated column "Movie Guide." "William Wilberforce and
his friends in the Clapham Sect brought faith and values
back to England in the early 1800s. They did so by
sounding the trumpet and conducting one of the most
effective commu-nications campaigns ever, yet there were
only a handful of them."

Through their heroic efforts, England's parliament outlawed slavery all throughout the British Empire.

"Revival, reformation and renewal can come to America..." Baehr said. "To do so, we need a cultural reformation."

NBC commissioned a poll in 2007 that asked the question whether the words "In God We Trust" on our cur-
rency and God in the Pledge of Allegiance should be pre-
served.

The network reported it had the greatest response of
any poll ever taken.

An overwhelming 86 percent of the respondents were in
favor of keeping the words, 14 percent were opposed. I'm
sure the extreme liberal leadership at NBC was both
shocked and disappointed by the results.

We are a significant, but silent majority in this land.

Author Chuck Colson also believes Christians are facing
two major threats in America today: Islam in the East and

Western society's abandonment of a belief in absolute truth.[3] He shared his concerns during a speech to several thousand pastors during the Southern Baptists Pastors Conference.

"While Colson touched on the threat posed by Islam – a belief system he described as using conquest to advance its cause – he focused his address on the culture war that American Christians face at home," said Tim Ellsworth writing for *Baptist Press*.

Colson said that two-thirds of the American people no longer believe there is such a thing as moral truth and that leads to "moral decay."

He said that in the past, most atheists who did not believe in God kept quiet about their beliefs, but there is a new breed of atheists preaching that Christianity is dangerous to society and should be regulated by the government.

"To engage the culture and counter the prevailing belief that truth is relative, Colson said Christians must do better at explaining... what they believe... Christians must understand that Christianity is more than simply a personal relationship with Jesus..." Ellsworth reported.

We must make it quite clear that Christianity is a worldview, the pathway to ultimate truth, he said.

"What is wrong with us when kids are being raised to believe there is no such thing as truth?" Colson asked. "That's the end of the Christian Gospel if we can't make a truth claim in our culture today."

Colson also admonished the thousands of pastors gathered for the conference: "Do not despair... no matter what happens around us. Despair is a sin because it denies the sovereignty of God. All He's asking us to do is fight the good fight."

Author and syndicated columnist Janet Folger urges pro-family activists to take a tip from former President Ronald Reagan: "We win, they lose."

"'We are those fighting for life, liberty and family','" Folger quoted Reagan.

Folger reminds us that Reagan did not start out just to survive the Cold War but to win it.

"We have every right to dream heroic dreams," Reagan said during a time when there was little reason to hope, like today.[4]

Dr. Martin Luther King, Jr., once said history will record "that the greatest tragedy of this period of social transition, was not the vitriolic words and actions of the bad people – but the appalling silence and indifference of the good people."

"Our generation will not only have to repent for the words and acts of the children of darkness, but also for the fears and apathy of the children of light," King said.

A group of pro-family students, parents and community leaders decided to fight back when one Washington State high school planned a "Day of Silence" event promoting homosexuality in the school.

Although the Gay, Lesbian, Straight Education Network reportedly sponsored the event to draw attention to the persecution of students in the school who are homosexuals, Pastor Ken Hutcherson of the Antioch Bible Church in Redmond claims the event was designed only to promote the homosexual lifestyle.[5]

Reverend Hutcherson and his wife purchased a half-page advertisement in the local newspaper urging other parents to join them and protest the event.[6]

"Then the word went out," he explained and some 300 parents of the school district attended a protest meeting.[7]

They were successful as 638 kids stayed home from school on the "Day of Silence."

According to Jeff Johnson writing for *OneNewsNow*, only 200 of the school's 1,400 students actively participated in the full day of activities.

"Hutcherson considers his protest of the homosexual event a 'very big success,'" Johnson said.

That's a victory for our side!

Remember how the federal government banned all references to God during flag-folding ceremonies at military funerals? After complaints from veterans' groups and thousands of other concerned citizens, the government recanted and allowed volunteer honor guards to use any biblical text requested by family members for the ceremony.

Hooray for the veterans groups!

There was another outcry by outraged citizens when the government removed the words "*Laus Deo*" or "Praise be to God" from an inscription on the Washington Monument.

According to the American Family association, officials of the National Park Service have changed their mind and will display the words in such a manner that they will be visible to the public.

Good things happen when people of good will take a stand!

Even the Capitol architect in Washington decided to prohibit any reference to God in the certificates accompanying flags flown over the capitol and purchased by citizens. The phrases include "Under God," "God bless you," and "In the year of our Lord." These phrases have always been used in the certificates.

The architect said he removed the words because references to God and the Lord might offend some Americans but now has changed his mind and the phrases will be returned to the certificates.

Well, taking them away offended me! And I'm an American!

A vocal group of black protesters is fed up with music videos they say degrade women and black and Hispanic men.

They have been gathering outside the homes of Viacom executives in Washington and New York City wearing red T-shirts with red stop signs and chanting, "BET does not reflect me. MTV does not reflect me," the *New York Times* reported.[8]

Viacom owns BET, MTV and several other networks.

"A lot of rap isn't rap anymore, it's just people selling their souls," said Marc Newman from New Rochelle, N.Y.

The protests, which began outside the home of Debra L. Lee, chairman of BET, were led by a group known as Enough Is Enough. The rallies attracted hundreds from the Enough group, the National Congress of Black Women and several other civil rights groups.

Organizers say the programs on BET and MTV glamorize video vixens and foul-mouthed pimps and thugs.

Right on!

Hooray for Elisabeth Hasselbeck!

She is a star of *The View* and one of the only pro-life voices on national television. Christians everywhere should be proud of her.

Several high-profile celebrities including Rosie O'Donnell, Whoopi Goldberg and Barry Manilow have criticized her unswerving defense of the rights of the unborn.

When O'Donnell was the host of *The View* on ABC, she often criticized Hasselbeck for her conservative views on abortion, the war on terror and her defense of the Christian faith. However, Hasselbeck never backed down.

After O'Donnell left the program in a huff, ABC replaced her with Goldberg, another strong abortion advocate.

Everyone knew that in time Goldberg and Hasselbeck would clash.

It started when Hasselbeck mentioned that she approved of one of the presidential candidate's proposal to give \$5,000 to the mothers of each new baby. She explained it might reduce abortions.

Goldberg went ballistic and said, "Back off a little bit… very few people want to have abortions."

Say what? According to Goldberg's biographer, she has had six abortions.

Peter J. Smith with *LifeSiteNews* noted that those "very few people" have killed nearly fifty million unborn children since Roe v. Wade was legalized in 1973.[9]

Goldberg also told Hasselbeck that the American people should revere women who have killed their unborn. Then she repeated the worn-thin mantra that Roe was necessary to make abortions "safe" and "clean" because women were dying from coat hanger abortions.

But most all surveys reveal that 97 percent of abortions are for birth control and/or convenience.

When singer Barry Manilow was scheduled to appear as guest on *The View*, he declined saying he believed Hasselbeck's views were "dangerous" and offensive." He added that although he was "a big supporter of the show," he must cancel his appearance because he didn't want to compromise his personal beliefs.

Isn't it interesting how O'Donnell, Goldberg and Manilow call evil good and good evil. Seems like I read something about that in the Bible.

Some believe that "Christian bashing" like the attacks on Hasselbeck is the last acceptable form of bigotry in this land and it is growing exponentially every day.

Now it's time for Christians everywhere to challenge this bigotry that is like a poison in our culture.

Officials of the Brentwood, California, schools cancelled a cross-dressing day after a large number of parents voiced their complaints.

"Recently, students… were encouraged to dress like the opposite sex during the last day of the school's 'Spirit Week.'" said Allie Martin writing for *OneNewsNow*. "The mother of a seventh-grader found out about the activity and contacted the principal to express her concerns."

She then contacted the Pacific Justice Institute and the legal organization coached her and other parents how to fight the school's cross-dressing plans.

"The only purpose that seemed to be involved with this event was for the sensitivity and tolerance of cross dressers, transsexuals and transvestites," said Brad Dacus of the Institute. "That's what the school was trying to push on these young girls and boys at junior-high age."

The school's principal said all he wanted was to encourage students to be "free thinkers" and that is why he chose to have the gender switch day.

Let's hear it for the parents who took a stand and won an important battle!

Here's some more good news in the "David v. Goliath" battle.

The Missouri Legislature passed a new law that forbids Planned Parenthood to teach sex education classes in the public schools of that state.

"The legislation's enactment gave permanent status to the Missouri Alternatives to Abortion Services Program, allowing schools to emphasize abstinence during sex education classes," said Jennifer Thurman writing for *Baptist Press*. "The law also prevents any personnel affiliated with abortion clinics from teaching sex education classes."[10]

Missouri Governor Matt Blunt who signed the bill into law said, "All life is precious and needs to be treated with the utmost dignity and respect."

"Abortion providers like Planned Parenthood should not be supplying our students with information about sexual health," the governor said. "This vital legislation ensures that our children get the information they need from their teachers, parents and physicians."[11]

Hooray for our side!

Wouldn't it be wonderful if Christians and conservatives in states throughout mid-America, the south and southwest would say "yes" to abstinence and life and "no" to Planned Parenthood?

Author and columnist Tristan Emmanuel says it's time for Christians and conservatives to stop apologizing to the secular progressives and take a stand for what we believe.

"With all the apologizing going on, I feel compelled to issue my own apology," Emmanuel said.

"I'm sorry that too many Christians can't discern between meekness and weakness.

"I'm sorry that we have confused virtues like compassion with inferior motivations like cowardice...

"I'm sorry that some 'conservatives' feel they need to pile on the condemnation of Ann Coulter simply because they don't like her style, or because she had the courage to say what the Church has been saying for more than 2,000 years."

Emmanuel says he is convinced "the secularists aren't winning the battle" – "Christians are... losing the battle."

"Christians are the ones lying down," he said. "Christians are the ones responsible for surrendering.

"And Christians are guilty of shooting their own kind, the few that are fighting the good fight."

Remember the marvelous movie *Facing the Giants?*

The movie, produced by the Sherwood Baptist Church in Albany, Georgia, has been shown in 57 countries and on every continent.

According to the Reverend Michael Catt, senior pastor at Sherwood, he has heard from people in Russia, Iraq, Australia, South America and China who have seen the movie.

Catt also reported that the movie was shown on the Disney Cruise Line and Turkish Airlines.

"The overwhelming majority of responses have been positive," he said. In the more than 10,000 e-mails received "people have thanked us for making the movie, told us of answered prayers and given us their story of how God used the movie in their own spiritual pilgrimage."

He added that more than 3,000 people trusted Christ as Savior as a result of the movie.

There is even some good news out of Harvard.

Religion has made a dramatic resurgence at Harvard, according to Harvey Cox, professor of religion at the nation's most prestigious school.

"In the early 1980s the faculty asked him to teach a required course on Jesus in the... undergraduate curriculum," said Gary Bergel in the *Intercessors for American Newsletter*. "After a couple of years, 700-800 students (including Jews, Muslims, and Hindus) were signing up to take the Jesus course."

Cox said that the president of Harvard invited him to lunch to try to understand the new religious phenomenon at the school.

"Jesus is much larger than the Church or Christianity," Cox explained to him. "Jesus is a very, very popular and interesting figure across the board."

Cox documented the religious resurgence in 2004 in his book *When Jesus Came to Harvard: Making Moral Choices Today.*

"Students are filling churches all around Harvard's campus," Bergel said.

Cox said the vast majority of the Harvard faculty believe "it's really irresponsible to turn students out into the modern world for leadership positions and government, business, media, whatever, without their knowing something about religion."[12]

Christians everywhere should be proud of Tony Dungy, the celebrated former coach of the Indianapolis Colts football team.

Dungy was the guest speaker as the Athletes in Action,

a ministry of Campus Crusade for Christ, celebrated the 19th Annual Super Bowl Breakfast in Miami, Florida.

Some 2,500 people in attendance greeted Dungy with several standing ovations before media from throughout the world.

"I'm very proud to be the first African American head coach in the Super Bowl along with my friend Lovie Smith," Dungy said, "but more than that the fact you have two Christian coaches… You have coaches who have firm Christian values, and the country and the world need to see that this week."

Dungy said being a football coach gave him a wonderful platform to witness for Christ.

"People know we're excited to be in the Super Bowl, but this is so small compared to what God has for me," he said.

Lovie Smith, coach of the opposing Chicago Bears, agreed with his friend Dungy in a video.

"God has given us a perfect stage to confess our faith in Jesus Christ," Smith said. Faith calls men "to be strong in Christ and successful in Him. Every day I start the day with God's Word…"

At the close of his remarks, Dungy invited anyone present who had never received Christ into their hearts to do so that morning.

Praise the Lord for two courageous culture warriors like Dungy and Smith.

By the way, the Colts defeated the Bears 29-17.

Some 25,000 teenagers attended an Evangelical Christian rock concert in San Francisco where they heard Christian rock music and strong warnings about drugs and immorality.

"It was part of 'Battlecry For a Generation,' an event to encourage evangelical Christian youth to fight back against a pervasive popular culture they say promotes sex, violence, drugs and alcohol," said Janine DeFao, writing for the *San Francisco Chronicle.*

The newspaper reported that similar events drew 7,000 youth to the Cow Palace in Los Angeles and 9,000 to Sacramento's Arco Arena.

"We're in the middle of a spiritual battle but also in the middle of a cultural war," Battlecry leader Ron Luce told the throngs of young people. "Your generation is being pounded with sexual messages... All the messages being sent through the movies, the Internet, point-and-click pornography. It's literally destroying your generation."

Luce is the founder of Teen Mania, sponsor of the event.

I'm sure he chose San Francisco for a good reason but I wouldn't set foot in that city even to see Adam and Eve.

Hooray for Teen Mania!

Author and psychologist James Dobson of Focus on the Family has thrown down the gauntlet and challenged this generation to listen to the call to change our land.

"Who will defend the unborn child in years to come?" he asks. "Who will speak for the aged who are no longer productive? Who will plead for the Terri Schiavos of the world who can be starved to death legally for having the misfortune of being disabled? Who's going to fight for the institution of marriage, which is on the ropes today? Who will teach young people the dangers and evils of both heterosexual and homosexual promiscuity?"[13]

Dobson says he is praying that the Lord will anoint a new generation of courageous men and women who will take a stand for the defense of righteousness.

My prayer is that it will not be too little, too late. Only time will tell.

Endnotes

Chapter One
1. David Kupelian, *The Marketing of Evil*, WorldNetDaily Books, 2005.
2. Tom Tancredo, *In Mortal Danger: The Battle for America's Borders and Security,*
WorldNetDaily Books, 2006.
3. William J. Federer, *Back Fired,* WorldNetDaily Books.
4. Jim Nelson Black, *When Nations Die* (Wheaton: Tyndale House, 1994), 187.
5. Thomas Sowell, "Taking America for Granted," TownHall.com, July 4, 2007.
6. Ibid.
7. Lee Iacocca with Catherine Whitney, "Where Have All the Leaders Gone?" (New York: Scribner), 2006.

Chapter Two
1. New Worldview, "Absolute Truth." AllAboutPhilosophy.org.
2. Kelly Boggs, "Ominous Diagnosis remains true," *The Christian Index,* Oct. 21, 2004.
3. Charles R. Kesler, "Limited Government: Are the Good Times Really Over?"
Imprimis, March 2008, A1.
4. Ibid.

Chapter Three
1. David Roach, Report on R. Albert Mohler, Jr. Interview on Cable News Network, *Baptist Press,* Aug. 21, 2007.
2. Pat Boone, "Farewell to the family?" WorldNetDaily Column, March 29, 2008.
3. Sebastian Ruiz, "Crowd attacks police during Labor Day beach brawl," *Beach and Bay Press,* Sept. 4, 2007.
4. Ed Thomas, "Suicide, Morality Collide on College Campuses," *WorldNewsNow,* Oct. 13, 2007.
5. Kelly Boggs, A Commentary, *Baptist Press,* Nov. 2, 2007.
6. Penna Dexter, "The burgeoning modesty movement," *Baptist Press,*

Sept. 27, 2007.

7. Warren Wiersbe, *Be Worshipful Glorifying God for Who He Is,"* (Colorado Springs: David C. Cook, 2004), 29.

Chapter Four

1. Joseph Farah, "The day socialism comes to America," WorldNetDaily, Feb. 17, 2008.

2. Sara Bonisteel, "Canada's Expecting Moms Heading to U. S. to Deliver," *Fox News,* Oct. 10, 2007.

3. "Top shrink concludes liberals clinically nuts," WorldNetDaily, Feb. 15, 2008.

4. Ibid.

5. "The War on Talk Radio," *Whistleblower Magazine,* July 30, 2007.

Chapter Five

1. Dick Morris and Eileen McGann, "Do-Nothing Congress – Big Salary, Little Work, Fee Trips," TownHall.com, June 23, 2008.

2. Ibid.

3. Pablo Guzman, "On Your DIME: Congressmen Lease Luxury Cars," CBS News, May 1, 2008.

4. Ernest Istook Guest Column, "Blame Congress for soaring gas prices," WorldNetDaily, April 24, 2008.

5. Bob Unruh, WorldNetDaily, Dec. 12, 2007.

6. Ellis Washington, The Report from Washington, "California Burning," Oct. 27, 2007.

7. Frederic J. Frommer, "Design Flaw Cited in Bridge Collapse," Associated Press, Jan. 15, 2008.

8. Jeff Poor, "Kerry Blames Tornadoes on Global Warming," Business and Media Institute, Feb. 6, 2008.

9. Ben Shapiro, "The Death of Shame," WorldNetDaily, March 12, 2008.

Chapter Six

1. Thomas Frank, "Busy Airports Miss Fake Bombs in Test," *USA Today,* Oct. 18, 2007.

2. Ibid.

3. Michael Hoffman, "Nuclear warheads mistakenly flown on B-52, landing at Barksdale Air Force Base," *The Military Times,* Sept. 4, 2007.

4. Fox News, Feb. 7, 2007.

5. Marcia Dunn, "Sabotage, Drinking Reports Shake NASA," Associated Press, July 27, 2006.

6. Ibid.

7. John Barry, *Newsweek,* July 5, 2007.

8. Ibid.

9. Ryan J. Foley, "Imprisoned sex offenders get thousands in federal grants," Associated Press, March 17, 2008.

10. Ibid.

11. Dr. Rachel Ehrenfeld and Alyssa A. Lappen, "US Funds Palestinian Terrorism," The Terror Finance Blog, March 2008.

12. Ibid.

13. Ibid.

14. Jesse Lee Peterson, "Moral poverty costs blacks in New Orleans," WorldNetDaily, Sept. 21, 2005.

15. Thomas A. Schatz, "Ideas on Liberty, Foundation for Economic Education," *The Freeman,* June 1996.

Chapter Seven

1. David Kupelian, "Why Hollywood is insane," *Whistleblower Magazine,* November 2006.

2. Will Hall, "Kathy Griffin is on the wrong list," Baptist Press, Sept. 14, 2007.

3. Ibid.

4. Robert S. McGee and Caryl Matrisciana, "Harry Potter: Witchcraft Repackaged," DVD.

5. Ed Thomas, "Hollywood misses mark again with TNT's *Saving Grace,*" OneNewsNow, Aug. 8, 2007.

6. Ibid.

7. "Abstinence advocate calls on CBS to apologize,"" WorldNetDaily, Oct. 6, 2007.

8. Bret Prelutsky, "Where have all the moguls gone?" WorldNetDaily, May 21, 2008.

9. Jim Brown, "Campaign targets MTV's, BET's music video 'assault' on children," OneNewsNow, April 11, 2008.

10. Ibid.

11. Michael Bates, *Reporter* Newspapers, March 15, 2007.

12. Ibid.

13. Ibid.

14. David Kupelian, "The secret curse of Hollywood stars," WorldNetDaily, Feb. 20, 2007.

15. Bret Prelutsky Column, "Hollywood's derriere-kissing sycophant," Dec. 5, 2007.

16. Phil Boatwright, "Christ-phobia in the movies," Baptist Press, Sept. 5, 2007.

17. Floyd and Mary Beth Brown, "Hollywood's revisionist history," WorldNetDaily, Jan. 26, 2007.

18.Warren Allen Smith, "Celebrities in Hell…"

19. Brian Fitzpatrick, "No forgiving Charlton Heston," *Perspectives,* April 8, 2008.

Chapter Eight
1. Dan Barry, David Barston, Jonathan D. Glater, Adam Litpak and Jacques Sternberg, *The New York Times,* May 11, 2003.
2. David A. Maraniss, *The Washington Post,* April 16, 1981.
3. Ibid.
4. Warren Wiersbe, "Be Worshipful Gorifying God for Who He Is (Colorado Springs: David C. Cook, 2004) 29.

Chapter Nine
1. Bob Unruh, "31,000 scientists reject 'global warming' agenda," WorldNetDaily, May 19, 2008.
2. Ibid.
3. Ibid.
4. Ibid.
5. "Sizzling study concludes: Global Warming 'hot air,'" WorldNetDaily, Aug. 20, 2007.
6. Phil Brennan, "Global Warming? New Data Shows Ice Is Back," Newmax.com, Feb. 19, 2008.
7. Ibid.
8. Richard Harris, "The Mystery of Global Warming's Missing Heat," National Public Radio, March 20, 2008.
9. Jim Brown, "British policy advisor says Gore is in 'panic mode,'" OneNewsNow, April 1, 2008.
10. Movie Trailer from Paramount Classics, a Division of Paramount Movies, 2006.
11. Mike Morris, "Billionaire environmentalist says world has too many people," *The Atlanta Constitution-Journal,* April 3, 2008.
12. Fred L. Smith, Jr., "Sick and Tired of Gore," Newsmax.com, March 3, 2008.
13. Ibid.
14. Paul Rogers, "Bill would require California's science curriculum to cover climate changes," *Mercury News,* Feb. 16, 2008.
15. Ibid.
16. Steve Lytle "Gore gets a cold shoulder," *Sydney Morning Herald,* Oct. 14, 2007.
17. Wesley Pruden, *Jewish World Review,* April 13, 2007.
18. "Al Gore and NBC: Birds of a Feather," *Investor's Business Daily,* July 20, 2007.
19. "Hysteria: Exposing the secret agenda behind today's obsession with global warming," *Whistleblower Magazine,* March 1, 2007.

20. John McCaslin's Inside the Beltway Column, Aug. 14, 2007.

21. "Challenge to Scientific Consensus on Global Warming: Analysis Finds Hundreds of Scientists Have Published Evidence Concerning Man-Made Global Warming Fears," Hudson Institute, Sept. 12, 2007.

22. Ibid.

Chapter Ten

1. "Illegal immigration costs border counties millions," Associated Press, March 5, 2008.

2. Jim Brown, "Economist says cost of illegal immigration rivals federal deficit," OneNewsNow, Oct. 5, 2007.

3. Pamela Hughes, "Money for Illegals' Care Runs Out," KTAR.com, March 17, 2008.

4. Ibid.

5. Chad Groening, "Former deportation agent claims thousands of criminal aliens released in U.S.," OneNewsNow, Oct. 16, 2007.

6. Ibid.

7. "Mexican military, U. S. police have border standoff in Texas," WorldNetDaily, Oct. 17, 2007.

8. "Battle on the Border Reaches to the Tri-State," Americans for Legal Immigration, November 2007.

9. "Mexico's Calderon protests U. S. crackdown on immigrants," Reuters, Sept. 2, 2007.

10. Ibid.

11. J. Michael Waller, "Why don't we try Mexico's immigration law?" *The Providence Journal,* April 29, 2006.

12. Ibid.

Chapter Eleven

1. Sarah-Kate Templeton, "Royal College warns abortions can lead to mental illness," *The Sunday Times,* March 16, 2008.

2. Ibid.

3. Ibid.

4. Ibid.

5. Janet Folger Column, "It's Huckabee or Hillary," Oct. 9, 2007.

6. Rusty Pugh, "Abortion #1 killer of blacks, says Tennessee pastor," OneNewsNow, Aug. 6, 2007.

7. Julie Rover, "New OB/GYN Guidelines Stir Ethics, Legal Debate," National Public Radio, March 20, 2008.

8. Ibid.

9. "Young Americans Hold Conservative Views," Focus on the Family, June 28, 2007.

10. Ibid.

Chapter Twelve

1. Hugh Miles, "Inconvenient Truths," *London Review of Books*, June 21, 2007.

2. Ibid.

3. Richard Esposito and Jim Sciutto, "U. K. Terror Plot – Why the Bombs Failed," ABC News, July 3, 2007.

4. "Report: Non-Muslims Deserve to be Punished," Fox News, April 1, 2008.

5. *The Daily Telegraph*, "Columbia seizes 60 lb of depleted uranium," March 28, 2008.

6. "Jihad USA: Confronting the Threat of Homegrown Terror," Fox News, March 27, 2008.

7. "Study: 3 in 4 U. S. mosques preach anti-West extremism," WorldNetDaily, Feb. 23, 2008.

8. "Imams Promote 'Our Values' on Taxpayer Dime," WorldNetDaily, March 15, 2008.

9. Charlotte Gill and Sam Greenhill, "Muslims planned to kill thousands by bringing down SEVEN transatlantic airlines in one go with liquid bombs," *Mail Online*, March 4, 2008.

10. Ibid.

11. Chad Groening and Jody Brown, "Cal Thomas criticized, commended for telling it like it is about Islam," OneNewsNow, July 7, 2007.

12. Jim Brown, "American, British religious freedoms on parallel tracks, says barrister," OneNewsNow, Jan. 28, 2008.

13. "Archbishop of Cantebury: Sharia law unavoidable in Britain," *Christian Today*, Feb. 7, 2008.

14. Sean Poulter and Niall Firth, "Bishop says collapse of Christianity is wrecking British society – and Islam is filling the void," *The Daily Mail*, May 29, 2008.

Chapter Thirteen

1. Toni-Lynn Robbins, *Bangor Daily News*, Nov. 3, 2007.

2. Ibid.

3. "Prof to student: Keep the faith, lose the grade," WorldNetDaily, April 11, 2008.

4. Chad Groening, OneNewsNow, Nov. 5, 2007.

5. "University of Vermont adds gender-neutral bathrooms," Associated Press, Aug. 26, 2007.

6. Vince Haley, "Save the Wren Chapel," *National Review*, Nov. 17, 2006.

7. "Sex show returns to college that banished cross," WorldNetDaily,

Jan. 31, 2008.

8. "Ward Churchill," DiscoverTheNetwork.org., Feb. 16, 2005.

9. Ibid.

10. Ibid.

11. Star Parker Column, "A papal message about academic freedom," April 19, 2008.

12. Chad Groening, "Harvard affords special 'exercise' privileges to Muslim women," OneNewsNow, March 7, 2008.

13. "Phyllis Schlafly calls protestors 'a bunch of losers,'" Associated Press, May 14, 2008.

14. Kavita Kumar, "Hundreds turn back on Schlafly at ceremony," *St. Louis Post-Dispatch*, May 16, 2008.

15. "Firing raises 1st Amendment, equal protection issues," WorldNetDaily, May 13, 2008.

16. Stephen C. Mayer, Access Research Network, Nov. 24, 1993.

17. Ibid.

Chapter Fourteen

1. Yvonne S. Waite, "Oprah into Spiritism," Bible for Today, Collingswood, NJ, Feb. 6, 1999.

2. Marilyn Ferguson, *The Aquarian Conspiracy* (Los Angeles: J. P. Tarcher, 1980), 23.

3. Randall Baer, *Inside the New Age Nightmare* (Lafayette: Huntington House, Inc., 1989) v.

4. Ibid.

5. Ibid.

6. Russ Wise, "Oprah Winfrey: the Oprahfication of America," Christian Information Ministries, Nov. 23, 2005.

7. Eric Butterworth, *Discover the Power within You* (New York: Harper and Row, 1968), 28.

8. Deepak Chopra, *The Seven Laws of Spiritual Success* (San Rafael: Amber-Allen Publishing and New World Library, 1994), 3

9. Kelly Boggs, "Chopra's 'third' Jesus is a fake," Baptist Press, March 28, 2008.

10. LaTonya Taylor, "The Church of O," *Christianity Today*, April 1, 2002.

11. Steve Cable, "Oprah's Spirituality: Exploring 'A New Earth,'" Probe Ministries, 2008.

12. Jeff Johnson, "Oprah labeled 'false prophet,'" OneNewsNow, March 27, 2008.

13. Penna Dexter, "Oprah's New Age Gospel," Baptist Press, March 27, 2008.

14. R. Albert Mohler, Jr., "The Church of Oprah Winfrey – A New

American Religion," Mohler's Website, Nov. 29, 2005.

Chapter Fifteen
1. William J. Federer, "Tyranny of the atheist minority,"
WorldNetDaily, May 16, 2008.
2. Ibid.
3. Ibid.
4. Ibid.
5. Ibid.
6. Chad Groening, "Pentagon generals in hot water for helping
promote Christian ministry," *OneNewsNow,* Aug. 24, 2007.
7. Michael Medved, "Hitchens Vs. God," TownHall.com, July 11,
2007.
8. Peter T. Chattaway, "The Chronicles of Atheism," *Christianity
Today,* Dec. 10, 2007, 36-37.
9. Ibid.
10. Warren Allen Smith, *Celebrities in Hell: A Guide to Hollywood's
Atheists, Agnostics, Skeptics, Free Thinkers, and More* (Fort Lee: Barricade
Books, Inc., 2002).
11. Bill O'Reilly, "Beyond belief," *Jewish World Review,* June 11, 2007.
12. The Connecticut Valley Atheists of Vernon, Connecticut, erected a
10foot, 3-sided sign that said "Imagine no religion?" WorldNetDaily,
Dec. 9, 2007.
13. Chuck Colson, "A New Breed of Atheist," *Christianity Today,* Aug.
2, 2007.
14. Ibid.
15. Jonathan Petre, *Weekly Telegraph,* London, Jan. 2, 2008.
16. Ibid.
17. Ibid.
18. Colleen Raezler, *LifeSiteNews,* Dec. 14, 2007.
19. Allie Martin, *OneNewsNow,* July 22, 200
20. American Family Association, Action Alert! Nov. 8, 2007
21. Kristen Fyfe, Culture and Media Institute, Dec. 21, 2007
22. Ibid.
23. Ibid.

Chapter Sixteen
1. Erin Roach, "Teens at risk as permissive sex yields STD epidemic,"
Baptist Press, March 14, 2008.
2. Christina Quick, "More College Students Walking Away from
Christianity," *Christian News Today,* May 19, 2008.
3. Mark Kelly, "Unchurched less likely to come to church," Baptist
Press, April 23, 2008.

4. Eric Gorski, "Survey Finds Religious Landscape in Flux," Associated Press, Feb. 25, 2008.

5. Allie Martin, "'Earth Worship' on the rise among evangelical youth," OneNewsNow, Feb. 25, 2008.

6. Tristan Emmanuel, "How the church has emasculated men," WorldNetDaily, Jan. 31, 2008.

7. J. Lee Grady's Fire in my Bones Column, "Desperate Times," *Charisma Magazine,* January 2008.

8. James A. Smith, "Paul, Peter and moderate Baptists," Baptist Press, Feb. 29, 2008.

9. Ibid.

10. Ibid.

11. Allie Martin and Jodie Brown, "Conservative Episcopalian groups say priest's dual faith causes 'confusion,'" OneNewsNow, June 21, 2007.

12. Ibid.

13. David Kyle Foster, "The Judas Church," *Charisma Magazine,* September 2007.

14. Allie Martin and Jodie Brown, OneNewsNow, Oct. 26, 2007.

15. Chad Groening, "Ahmadinejad has 'blossoming friendships' with U.S. religious left, says Christian activist," OneNewsNow, Oct. 16, 2007.

16. Chad Groening, "National Council of Churches praises Ahmadinejad," OneNewsNow, Oct. 8, 2007.

17. Melanie Hunter-Omar, "Maher to Apologize for Calling Pope a 'Nazi,'" *CNS News,* April 17, 2008.

18. Wesley Pruden, "Nobody's pickin a church fight," *Washington Times,* July 13, 2007.

19. Phil Fernandes, "Approaching the 21st Century: The Death of God, Truth, Morality and Man," Evangelical Theological Society, Feb. 28, 1998.

20. Francis Schaeffer, *Complete Works, Vol. V* (Westchester: Crossway Books, 1982), 381.

Chapter Seventeen
1. David Kupelian, "Open letter to WND readers from David Kupelian," WorldNetDaily, Dec. 11, 2007.

2. Ibid.

3. Tim Ellsworth, "Engage the culture, Colson exhorts Baptists," Baptist Press, June 11, 2007.

4. Janet Folger Column, "Time to drain the cultural swamp," May 6, 2008.

5. Jeff Johnson, "'Day of Silence' walkout a success," OneNewsNow,

May 7, 2008.

6. Ibid.

7. Ibid.

8. Felicia R. Lee, *The New York Times,* Nov. 5, 2007.

9. Peter J. Smith, *LifeSiteNews,* Oct. 4, 2007.

10. Jennifer Thurman, "Mo. Schools Expel Planned Parenthood," Baptist Press, Oct. 11, 2007.

11. Ibid.

12. Gary Bergel, Intercessors for America *Newsletter,* December 2007.

13. James Dobson, "Who will answer the call?" Focus on the Family, Sept. 17, 2007.

Bill Keith is an award-winning journalist who served as an investigative reporter, city editor and editor of three newspapers in Louisiana and Texas. He was a war correspondent in Vietnam and also had assignments in Tokyo, the Philippines and West Berlin and traveled in 35 other countries.

He earned the bachelor of arts degree in writing / journalism from Wheaton College in Illinois, the master of divinity from the Southwestern Baptist Theological Seminary in Fort Worth, Texas, and received a graduate diploma from the Tokyo School of the Japanese Language.

Through the years he served as the director of public relations for the Baptist General Convention of Texas; a Louisiana state senator representing the people of Shreveport; senior editor of Huntington House Books, Inc.; and president and chief fundraiser for the Academic Freedom Legal Defense Fund.

He and his wife Vivian Marie live in Longview, Texas, and have six children in their blended family: Tara Rose, Kim Westfall, Richard Keith, Miguel Mendez, Marisa Murphy and Lindsay LeBell. The Keiths are members of Pathway Church in Longview.

Keith has written 20 books – both fiction and non-fiction – including *Days of Anguish, Days of Hope* (Doubleday); *The Commissioner* (Pelican Publishing Co.); *Scopes II/the Great Debate* and *The Divine Connection* (Huntington House, Inc.); *Joy Comes in the Morning* (MV Press); *W. A. Criswell/the Authorized Biography* (Fleming Revell); *Gettin Old Ain't for Sissies; The Prayer Bag and Other Stories that Warm the Heart;* and *The Magic Bullet* – a novel – all available from www.-BillKeithBooks.com.

Other Books by Bill Keith

The Prayer Bag and Other Stories that Warm the Heart takes the reader on a spiritual journey through the prayer lives of some of the greatest Christians the world has ever known... a missionary in the jungles of Sumatra who prayed for a nail and found one; a survivor of the dread Ravensbruck Concentration Camp in Germany who prayed for a brutal guard who brutalized her sister while in the camp; an evangelist who witnessed to Emperor Hirohito of Japan after World War II; and a preacher who carried a cross around the world. "And you will read about my dear wife Vivian Marie who carries a prayer bag with her everywhere she goes." ~ The Author (Available at www.BillKeithBooks.com)

Joy Comes in the Morning is the true story of one of the greatest miracles of the Twentieth Century. Delores Winder, a Presbyterian lady, was an invalid for 19 years and was planning her funeral when God intervened in her life. She was completely healed during a United Methodist Church Conference on the Holy Spirit in Dallas, Texas, in 1975. Since that time she and her husband Bill have traveled throughout the world telling her amazing story. (Available at www.BillKeithBooks.com)

The Magic Bullet is a novel about a scientist who discovers the secret of life extension. However, the discovery creates all kinds of problems for him. A recluse billionaire in Chicago – who is dying – wants to find the secret and sends his men to kidnap the scientist. Also, the Chinese

government hires a New York City *Mafia don* to find the
scientist and learn the secret. The scientist hides out in the
Barataria swamps below New Orleans and joins a motor-
cycle gang en route to the biker's rally at Sturgis, South
Dakota, where he is captured by the billionaire's men and
taken to Chicago. The reader will laugh and cry but will
never forget the dramatic climax. (BillKeithBooks.com)

Gettin' Old Ain't for Sissies is a motivational/inspirational
book to help the baby boomers and older survive and enjoy the senior years. The thesis is: "Old age doesn't have to
be the end of the line. It can be a bright new beginning."
The book outlines the five things a person must do to live a
vigorous lifestyle into the 70s, 80s and even the 90s and
gives numerous examples of Older Champions.

The Commissioner is the intriguing true story of death and
deception and reveals a corrupt political battle during the
1970s that threatened Shreveport, Louisiana. The city's police commissioner – the most powerful lawman in the state
– was behind multiple scandals including racism, payoffs,
theft of city funds and tampering with a grand jury. He
may also have been involved in the murder of an advertising executive who was scheduled to testify against him
in court. (*Available at your local book store*)

Days of Anguish, Days of Hope is the heroic story of Chaplain Robert Preston Taylor who spent 42 months in Japanese prison camps during World War II. On Dec. 7, 1941,
the Japanese bombed Pearl Harbor and Manila, the Philippines, the next day. Taylor was caught in a maelstrom of
war and the every-day fight for survival. He ministered to
the fighting men on the front lines during the Battle for
Bataan and received the Silver Star for bravery. He en-
dured the Bataan Death March – where thousands of
American soldiers died – the Cabanatuan Prison Camp,

and the so-called "hell ships" that were bombed by American pilots who did not know the American prisoners were on board. During the nearly four years in prison, he faithfully ministered God's love to the other prisoners. After he was liberated, he returned home to learn that his wife Ione, who was told he had died on the "hell ships," had remarried. He decided to remain in the military and years later President John F. Kennedy named him Air Force Chief of Chaplains with the rank of major general.

*All books available at www.BillKeithBooks.com

Look for these exciting books in the future: The Guns of Winter (A novel about one man's war with Washington); W. A. Criswell/the Authorized Biography about the dynamic pastor of the First Baptist Church of Dallas, Texas, and a man believed by some to be the greatest preacher of the Twentieth Century; and *The Guns of Winter,* a Bill Keith novel about the treachery and intrigue of one man's war with Washington; and others.

Made in the USA
Charleston, SC
05 June 2011